Table of Contents

A Note from the Author

We have all at some point in our lives thought about it. From watching and reading about characters like Peter Pan, never growing older, or Dr. Frankenstein, desperately trying to understand how to bring back life, at some point nearly all people think about what we, humans are made of. Who are we? We all look different. Why do children look similar to their siblings or parents? Why do we grow older? How do we grow older? Why do we die? Is it possible to live forever?

Prologue

Peter Pan was confused about the book Wendy gave him. So many pages it had. He looked at each side of it, then the top and then the bottom of it. With Wendy now gone and grown up, he had to tell the Lost Boys stories of the outside world, as they were saddened with the fact that they could not come with her. All they wanted now, was to learn of that world and of what happened in the past. Peter did not know what to do anymore. He knew that the Lost Boys would want to fight Captain Hook and his pirates, but they would always refuse to do so. They wanted to learn, and read from the book that Wendy gave Peter: about DNA and Biology.

Lost Boys: C'mon, Peter. Read it to us. Wendybird left it for a reason here.

Peter Pan: Would it cheer you up if we read through it? Would you grow up enough to not grow up?

Lost Boys: Yes!

The Lost Boys huddled around Peter as he slumped into his fur-covered throne. Tinkerbell sat on her chair made out of lilacs and began to listen:

Introduction: the Birthday of the World

"What is life? This has always been wondered by humans throughout history, thinking about afterlifes, or how women are different from men. At some point in time, scientists were able to discover the true understanding of what creates life. At first, the way life was created was through natural weather. The beginning of evolution started in what is known as the Primordial Soup, created about 4.6 billion years ago (when the Earth was born). 600 years after Earth's own evolution, life on Earth began to form. But why Earth and not some other planet, like Jupiter or Neptune? Water. Water is the most essential ingredient for life on Earth. Water has the ability to dissolve inside other molecules, which is why all cells are filled with water. Due to this, the only place Earth could make its first life form was in the seas, oceans, and lakes."

One of the Lost Boys stood up. "Earth. . . that is where Wendy is from. We are from there too, no?"

Peter Pan: I guess so. If you do not believe in the fact that there is only one universe and you just wound up here. I am from Neverland and only from Neverland. On Earth, they age. They grow old. Here, you can never grow up! Never! Moving on, and let us not interrupt. We will talk about what we read at the end of each. . . part. When the new ideas start again. Got it?

The Lost Boys all nodded, not saying a word. Peter continued.

"99% of living organisms' materials use carbon, hydrogen, oxygen, and nitrogen. Due to this necessity, these molecules had to stick together to create the necessary basis of living organisms. With help from the Sun.

Earth was very different back then. The biggest difference was not having an ozone layer- the layer that protects us from burning up while living on Earth. Due to having no ozone layer, the Earth had terrible ruins on its surfaces because of the ultraviolet Sun rays. Due to the heat, volcanoes would always erupt, and the sun rays would keep the oceans' temperature at 80°C (176°F)! This huge amount of heat being taken on the Earth's oceans caused the ocean's salt water to evaporate, which created clouds, storms, and concluded the fact that there were constant lightning storms and lots of rain.

The kinetic energy given to the molecules (when temperature rises, kinetic energy increases within particles), causes small molecules to basically stick together to form even larger molecules, creating big molecules like Ammonia, Hydrogen, Methane, and Carbon Dioxide. These are known as "organic molecules". Humans are made out of them. They are the building blocks of all other living things as well. Other important molecules that were created in general were amino acids, sugar, and nucleotides. Due to heat, the molecules did not stop. Specifically, the nucleotides. The nucleotides started sticking to each other, creating Polynucleotides. Amino acids also did the same thing, creating polypeptides. The polynucleotides created nucleic acids, which as a result created DNA and RNA. The difference between DNA and RNA is that RNA molecules are short and a single strand of material, whilst DNA molecules have two strands twisted around each other in a double helix. Polypeptides created proteins. DNA, RNA, and proteins all perform important roles in the way all living things live.

It took a long time for anyone to understand what living things truly were made of and how they differentiated from each other, non-living things. So who were the first people to do so? How has what was created evolved? And why can't we reverse these processes to change/revive lives?

Peter: I cannot read this any more. All of these strange words-- amino acids, nucleotides, polynucleotides. Ew, disgusting-- all grown up words. The only one I know is sugar because it is what is inside the mangos and the fruit on this island. It is delicious. . . Tink, what shall I do? I cannot read this without understanding only half of what I am reading.

Tinkerbell: Peter, there is a thing called a glo-ssa-ry. It has all of the definitions of the words in this book. Just continue reading! Interesting to see what these humans think about when they grow older.

Section 1: What is Life?

What is Biology?

Peter continued reading, slowing down at the words that he did not know well.

"Biology used to be the study of natural history-- Animals and classifying them into groups. However, that changed when in 1665, Robert Hooke, an English polymath examined a dead cork under a microscope, in which he noticed little "chambers" which he called cells.

Robert Hooke, known as the "Renaissance Man" of the 17th century for the contributions he made to different parts in science-- Biology, Physics, and Astronomy. He was born in July 1635, on the Island of Wright, England. His parents served in the local churches. At the age of thirteen, after his father's death in 1648, Hooke was sent to London to study under the apprenticeship of painter Peter Lely. These studies ended quickly, as he later decided to study at London's Westminster School. In 1653, Hooke enrolled into Oxford's Christ Church College, in which he was an assistant to scientist Robert Boyle, famous philosopher. During his time studying there, he became friends with future architect Christopher Wren. In 1662, Hooke became the curator of experiments for the Royal Society of London. During that time, Hooke performed many experiments, and was recognized for what he made while looking through microscopes. He also created the 'Hooke's law' of elasticity. "The invention of the microscope led to the discovery of the cell by Hooke. While looking at cork, Hooke observed box-shaped structures, which he called "cells" as they reminded him of the cells, or rooms, in monasteries". Later, he became a professor of geometry at the Gresham College of London. After the Great London Fire in 1666, Hooke became a surveyor. He died in 1703.

In 1838, Matthias Schleiden, a German botanist, and Theodor Schwann, a German physician concluded that cells are the smallest components of all living organisms. This created the cell theory: *all cells are born from other cells.*

After Robert Hooke published the book *Micrographia,* as well as the discovery of cells, the discovery of classical cell theory was created. This was proposed by German Physician Theodor Schwann, who created three parts of the cell theory. Part one states that all organisms are made up of cells. The second part: cells are the basic units of life-- this was determined after Matthias Schleiden and Theodor Schwann's observations when comparing plant and animal cells (experiment performed in 1838). Part three stated that cells come from cells. Ever since the formation of the cell theory, technology has improved, which allows for more detailed observations that led to more discoveries about cells.This helped the creation of the modern cell theory, also sustained by three parts.

I: DNA is passed between cells during cell division

II: The cells of all organisms of similar species are chemically and structurally mostly the same.

III: Energy flow (on the biological level) occurs within the cells

Later, Darwin came up with the theory of evolution, which stated that organisms are not static, and undergo variations from generation to generation. The living organisms with the "best" adaptations to the environment have a better chance in survival. This concluded that the central principle of biological science is evolution of living things.

Biology

Bio= life

logy= to study

Biology= the study of life (living organisms)

Later, discoveries of heredity (performed by Mendel) were made. By the twentieth century, molecular science had already begun. However, biochemists still wondered, "Do the laws of science apply to both living and nonliving things?" or "There must be something besides matter, a life force that distinguishes living things from nonliving things" or "Organisms probably contain a soul in them or something". This perspective was known as Vitalism.

Erwin Schrodinger, an Austrian-Irish physicist and researcher in quantum theory, wrote the book, *What is Life?* He made multiple observations within his book, stating that life is made of matter and therefore can be explained in the language of matter. Basically, this translates to living things are substances that are breathing so therefore can be understood what it means to be living. Schrodinger also said that organisms feed on negative entropy-- when a system can exchange either heat or matter with its environment, a disorder decrease of that system occurs. Nature tends to move in the direction of disorder (towards entropy), however, living things create order by 'sucking' orderliness (negative entropy) from the environment. This idea allowed for biologists to adopt experimental methods of physicists: working with big molecules.

Erwin Schrodinger was born in August 1887 in Vienna. Schrodinger's prime interests were not only in science but also in "the severe logic" of grammar of German poetry. From 1906 to 1910, he studied at the University of Vienna. He was later an assistant of Franz Exner, Austrian Physicist. During the First World War, he served as an artillery officer. Schrodinger's greatest discovery was the wave equation, discovered during 1926. In 1933, he shared a Nobel Prize for his work in analyzing Bohr's orbit theory and about atomic spectra. As his health was declining, he retired and returned back to Vienna, in which he died in 1961.

Through these experiments, important research was found, including:

I: A group of molecular geneticists created a way to use bacteria and viruses as "guinea pigs in laboratories"

II: Biochemical geneticists discovered that each gene is responsible for the production of a single enzyme-- regulates the rate in which chemical reactions go on

III: X-ray crystallographers used X-rays to determine the structure of protein molecules from the crystals

During the meantime, the discovery of DNA occurred, with Frederich Meischer discovering that cell nuclei contain nucleic acid, now known as Deoxyribonucleic Acid, or DNA. In 1944, Oswald Avery, a Canadian-American Physician proved that DNA transmits genetic information.

Oswald Avery was a scientist working at the Rockefeller Institute for Medical Research starting from 1913. In the 1930s, he focused on the research on a bacterial species known as

Streptococcus pneumoniae, which causes many various diseases. In the 1940s, through this bacteria, he worked on an experiment, famously known as the Avery experiment, which proved that bacteria without capsules could be turned into bacteria with capsules by adding material from a capsulated strain. This discovery became known as the 'Transforming Principle' and through his experiments, Avery and his collaborators found that the transformation of the bacteria was due to DNA. This concluded the fact that DNA is the carrier of genetic information. Previously, scientists thought that traits were carried by proteins and that DNA was too simple to be the carrier of genetic information. Later, Avery's work was more focused on the capsule of different strains of the Streptococcus pneumoniae bacteria, because he thought the capsule was an important part in the disease that the bacteria caused. He even found out that the strains without a capsule were completely harmless. In 1928, scientist Frederick Griffith managed to produce a disease in mice by using a live non-capsulated strain. Through Griffith's work, Avery decided to figure out what was passing into harmless non encapsulated strains from the dead capsulated strain. In the 1940s, Avery and his collaborators replicated Griffith's experiment. They later purified that substance that was carrying the transformation. They discovered that only 0.01 micrograms was sufficient to transform their live cells into encapsulated cells. Avery later analyzed the characteristics of the transforming substances. He tested their chemical make-up (EX: phosphorus content), which is actually also present in DNA but less so in proteins. Another test followed this, with Avery checking the substance's ultraviolet light absorption characteristics. Both tests concluded that DNA is the transforming substance of heredity and not proteins. In 1944, Avery published a paper with his research and experiments of his and his coworkers' discoveries.

And then, in 1953, Biochemist and Biophysicist James Watson and Francis Crick discovered the structure of DNA containing four nucleotides that correspond with each other.

In later years, discoveries about how genes were expressed, how proteins are made from genes, as well as others were made. This led to the development in the method of recombinant DNA technology. Now, researchers can 'cut and paste' parts of the DNA molecule, which allows them to analyze big DNA molecules found in higher animals, and allows the ability to quickly decipher DNA sequences.

This also allowed for scientists to be able to have genes that produce humans have the ability to be inserted in bacteria to create genes produced substances. This led to the cloning of Dolly the Scottish Sheep, who was cloned from one mammary gland cell.

History of Genetics

Although genes have always been inside of us humans, and have formulated the physical traits of humans and animals as we are, only recently (a couple of centuries back) has the idea of genetics and inheritance actually been somewhat understood. There were many theories on the idea of inheritance, however, before the 19th century, all theories were later proved to be incorrect.

The Ancient Greeks developed the theory of "pangenesis", an idea that claimed that having sex involved the transfer of small body parts into the embryo, which will later be part of a future child. These small parts included hairs, veins, tendons, bones, and more. This idea was supported by Charles Darwin's (a british Naturalist) theory of pangenesis, stating that all organs are able to create a small particle (which he named "gemmules") that were assembled in sex organs, and were given into the embryo through sexual reproduction. Darwin also added that any change after birth of a human or animal, will be passed on to the next generation (Ex: someone had one finger amputated after birth, and so their son or daughter shall have one finger missing as a trait of theirs). Later, Darwin supported Jean Baptiste Lamarack's theory of inheritance, which in fact disproved the theory of pangenesis, stating that there was a process of natural selection: if any organism during its lifetime adapts in some way to its environment, those changes will be passed on to their offspring. Darwin believed that natural selection was the key

of human and animal evolution, *however*, he thought that natural selection was based on a variation of pangenesis.

This theory was proved wrong when August Weismann, a German biologist, decided to perform an experiment by cutting off the tails of some mice, to see if that change would be seen in the mice's offspring. According to Darwin's theory, mice in the future generations would receive "gemmules" that say 'no tails' and the offspring would not have any tails anymore. However, this was proved wrong because Weismann's mice's future generations still had tails. In conclusion, the theory of pangenesis that Darwin and the Greeks thought of was proved incorrect.

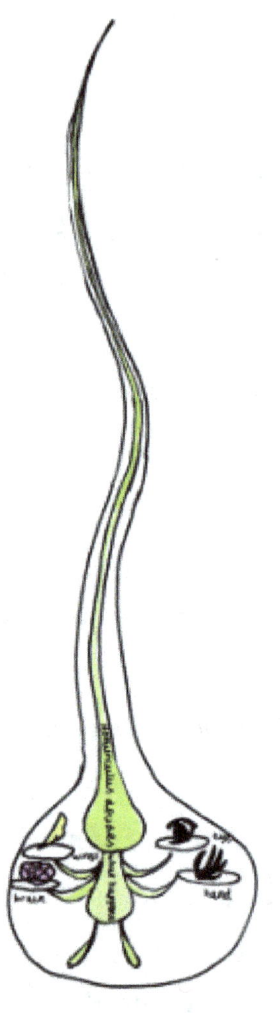

In the 17th to 18th century, another theory was created: preformationism. Preformationism is the idea that either a sperm or egg cell (nobody actually knew which one it really was) contained a little person called the homunculus, who would control what an offspring would look like. This theory was always questioned on how genetic disease would work based off of this theory:

I.   A manifestation of God, creating mischief (or thought of as devil/demon work).

II.   The result of unkind thoughts of the mother during her pregnancy

Thoughts from Peter Pan: I have two questions about this-- Why? Who thought of this stupidity? And to think grownups thought of this?

This theory was proved false, when in the early 19th century, as microscopes were slowly established, it could be clearly seen that there are no little humans living within a sperm or egg cell.

The theories all came to an end an Augustinian monk understood the idea of inheritance through growing little pea plants and his home.

His name was Gregor Mendel.
Known as the father of genetics.

Mendel

Gregor Mendel was born in the year 1822, to a farmhouse in the current Czech Republic. He excelled in school, however, he became a monk in an isolated abbey in Brno, Czec Republic. He failed at becoming a parish priest, and so wanted to become a teacher. However, to become a teacher, he had to take an exam, which he failed. Due to this, one of his father superiors took him to study at the University of Vienna, where he excelled in physics. However, after finishing university he still failed the exam. He rose to no more than the rank of a substitute teacher.

He took up the hobby of gardening, and became a botanist. He planted green peas and saw how they changed throughout the generations. Through this, he realized the idea of inheritance, and therefore created the theory of inheritance, in which there were multiple ideas:

I. Rule 1: Unit factors (genes) exist in pairs
II. Rule 2: When unit factors are unlike (different), one dominates the other (due to being in pairs)-- this is now known as a dominant and a recessive trait
III. Rule 3: Features are discovered/ inherited independently

Mendel later published his work, however nobody noticed it and was underappreciated. Only a few decades later in 1901, was his work understood to be ingenious, naming him the *father of genetics.*

Through Mendel's work on heredity, scientists were able to understand the idea of cells and how there is something inside of them that carry your genetic information. Scientists understood that this thing in the cell was what created life and what made us differentiate from each other. This is what is known as DNA.

But who was the first person to ever think of such an idea?

Frederich Miescher

It began, in 1869, with the description of DNA being discovered by Swiss researcher Frederich

Miescher, a man that nearly nobody had ever heard of. Like Mendel, Miescher's discovery of DNA at first was unappreciated as nobody understood what his discovery meant. When DNA was starting to be understood as one of the most crucial parts of Biology, Miescher had already died. . . half a century earlier.

Overall, Miescher was interested in all types of sciences from a young age. He grew up in a family full of scientists, most of whom were professors in the University of Basel.

One of his most famous relatives was his uncle, Wilhelm His (who was famous for his work on cells and tissues during embryonic development and the discovery of neuroblasts-- an embryonic cell which nerves come from). Miescher finished medical school at age 23 and specialized in otology. He failed working as a doctor due to his partial hearing disabilities, which affected his ability to communicate with patients. Instead, he decided to make a career out of his true passion: theoretical origins of life.

To discover more about the "theoretical origins of life", Miescher wanted to determine chemical compositions of cells (Meaning- a cell's chemical template/outline) and the principles of life in a cell using lymphocytes (a white blood cell with one nucleus) to help him achieve this goal. At first, Miescher experimented on various types of simple proteins that make up leukocytes (colorless cells that circulate in blood fluids). However, due to the lack of equipment, he was unable to determine a definitive composition.

--

During these tests, Miescher noticed that there was a substance coming out of a liquid called pepsin. Miescher's experiment showed that when acid was added and dissolved due to the addition of alkali. That was the first time Miescher found natural DNA. According to Max Plank from the Institute of Developmental Biology, "[Miescher stated that] 'according to known histochemical facts, I had to ascribe such material to the nuclei' and he decided to examine the cells' nuclei more closely—a part of the cell about which very little was known at the time"

(Plank).

--

Miescher's initial description of DNA was that it was not the same as a protein: "In my experiments with low alkaline liquids, precipitates formed in the solutions after neutralization that could not be dissolved in water, acetic acid, highly diluted hydrochloric acid or in a salt solution, and therefore do not belong to any known type of protein." Miescher called what is now known as DNA, "nuclein". Miescher was later able to isolate "nuclein" multiple times, however, for some reason, he was nearly forgotten by the science world, until the 1950s.

Unfortunately, the significance of Miescher's work about DNA was not accepted by the majority of people at the time. Most scientists believed that complex proteins held genetic information. It was not until the 1940s when medical and biology researchers Oswald T. Avery, Colin MacLeod, Maclyn McCarthy, Al Hershey, and Marta Chase all presented and demonstrated simple experiments that proved that DNA was the carrier of genetic information.

When the idea of DNA was beginning to popularize, scientists were starting to wonder what was made up of DNA. Many scientists contributed to the discovery of what DNA is made up of, but the most famous are. . .

James Watson, Francis Crick, and Rosalind Franklin.

The Double Helix Structure and the Three Biologists

The structure of DNA is quite a simple idea, however, it took years to find out what it actually was. After people found out about Miescher's discovery of DNA, scientists were starting to wonder what made up the DNA, and most importantly, what held it together.

In the 1950s, James Watson and Francis Crick (with the help of Rosalind Franklin) were able to discover what is known as the double-helix structure-- the template that holds genetic information.

James Watson and Francis Crick (American and British molecular biologists) met at the Cavendish Laboratory located in the University of Cambridge. They both came to the conclusion that genes have to be made from DNA. They also agreed that understanding the structure of DNA was key to solving the idea of creating genes. And so their adventures began by finding what DNA's structure is.

Watson and Crick built a DNA model based on the work of American chemist, Linus Pauling, who used a model to determine the structure of proteins.

Watson and Crick knew that DNA's structure consisted of long chains of nucleotides (described later on in this section). They understood that the structure must be big enough to fit in more than one strand of nucleotides. So, they created their first model-- the triple helix structure.

The triple structure they created had a chain of sugar and phosphate in the inside of the structure, and the base (a substance that helps with reactions between DNA and other acids) on the outside. Watson and Crick showed this model to other scientists, who were looking at DNA through X-rays. They told Watson and Crick that their model was not even close to what they saw. Humiliated, Watson and Crick set aside their DNA research and went back to their previous studies.

A year after their failure, Watson and Crick heard that Linus Pauling, the man whose model they used to base off of, figured out the structure of DNA. Pauling's idea was that the structure of DNA consisted of the triple helix structure, however, the inside and outside of his model was basically the reverse of what Watson and Crick created. James Watson looked at X-rays pictures of DNA and noticed that molecules contained more water than their previous samples. He also noticed that the pictures clearly showed the helical shape of DNA. These photos were produced by English chemist and X-ray crystallographer, Rosalind Franklin, who was discredited for her pictures, as Watson stated in his book *DNA: The Secret of Life*. Through these pictures, Watson and Crick understood that DNA did indeed have a helical structure. From this, Watson was reminded of what scientists like Mendel had discovered: "Nature comes in pairs". Through this, Watson decided that DNA was structured through the double helix structure (which he built a model of), which ended up to be correct.

Watson and Crick still had to find out how the nucleotides work. Watson thought that the nucleotides should be split into 2 pairs (there are a total of four nucleotides in DNA), and those pairs could form into identical bases. Crick disagreed with this, saying that if they would assemble the DNA molecules with the pairs of the same base being connected to each other, the shape of the DNA would be completely strange and definitely incorrect.

Later while looking at the 4 nucleotide bases separately: Adenine (A), Thymine (T), Cytosine (C), and Guanine (G), Watson realized that each nucleotides has a complementary nucleotide due to having the same size and shape: e.g: Adenine (A) goes with Thymine (T) and Guanine (G) goes with Cytosine (C). This gave the double-helix form to be refined as to having two phosphate chains, which helped with a stable configuration that did not fluctuate the thickness of the model. These base pairs were called complementary base pairs.

Not only did Watson and Crick discover the entire structure of DNA, but they also were able to discover how genetic information was transmitted.

--

In 1962, Watson and Crick won the Nobel Prize "for their discoveries concerning the molecular structure of nucleic acids and its significance for information transfer in living material."

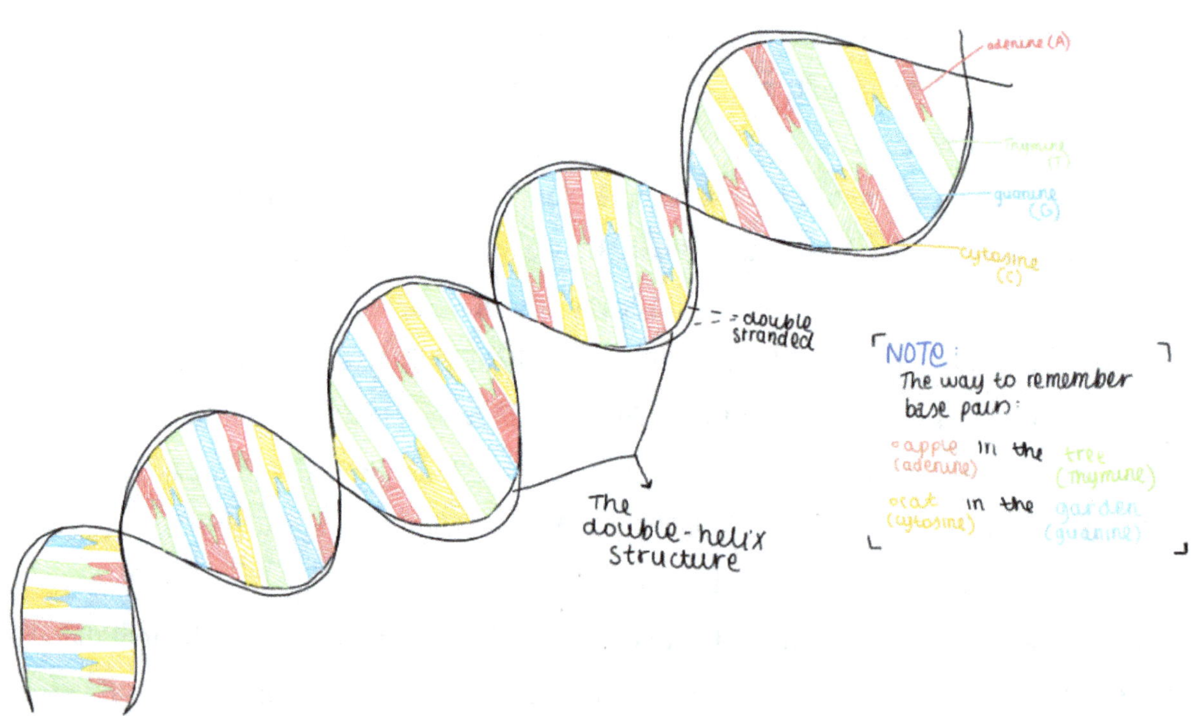

Rosalind Franklin was identified in helping with their discoveries, however even in partnership with Watson and Crick, did not receive the Nobel Prize, until her death at age 37.

--

Just like Watson and Crick's steps to discoveries, we can separate the shape of DNA into two parts: the double helix structure, and the four bases (which are called base pairs).

Double Helix Structure:
What exactly did Watson and Crick discover?

Double-helix structure

The double helix structure is what holds the genetic information in place. It is made of two individual strands which loop around each other antiparallel, and twist around each other (hence the double- helix part of the name). Each of those strands are made out of sugar and phosphate molecules (which give the nucleotides/base pairs to be able to attach to the structure). This "chain" binds together four nucleotides (also known as base pairs): Adenine (A), Guanine (G), Thymine (T), Cytosine (C). These base pairs are held inside the structure (so within the linkage of the two chains).

Base pairs

Nucleotides are molecules made of deoxyribose sugar (a sugar without water and a little molecule formula called Ribose), one phosphate group, and a nitrogen base. DNA has a total of four different types of nucleotides, which are located inside the double helix structure. These four nucleotides create two base pairs. This means that two certain nucleotides link together. Adenine links with Thymine, and Cytosine links with Guanine.

To remember which nucleotides base goes with which other one, remember:

Apple in the **T**ree   **A**denine and **T**hymine
Cat in the **G**arden   **C**ytosine and **G**uanine

These base pairs are what hold genetic information.
Base pair sequencing was discovered and established by Austro-Hungarian born American biochemist, Erwin Chargaff in 1949-1950s.

Erwin Chargaff was an American biochemist who discovered that DNA is the primary "constituent" of genes, impacting the approaches to studying the biology of heredity.

Chargaff was born in August 1905 in Chernivtsi, Austria, now Ukraine. He graduated from highschool at the Maximiliangymnasium in Vienna and later went to the University of Vienna. In 1928, he earned his doctoral degree in chemistry and later traveled to the US as a researcher at Yale University. He stayed in the US until 1930, in which he became an assistant in the public health department at the Berlin University in Germany. In 1933, Chargaff transferred to the Pasteur Institute in Paris. In 1935, he returned to the US to become an assistant professor of biochemistry at Columbia University. Seventeen years later, Chargaff became a full time professor there and became the chairman of the department from 1970 to 1974 as a retired (but allowed to keep his title as an honor) professor of biochemistry. Although Chargaff's primary occupation had to do with chemistry, his most important contributions were to the foundations of DNA.

During the time he was working, it was not known that genes were composed of DNA. It was thought instead that the 20 amino acids that compose proteins in a cell carried the genetic information. Scientists believed that due to the many kinds of amino acids which compose protein in the cell were able to be combined in enough different ways to form a sufficiently complex basis for a gene.

These beliefs ended in 1944 when O.T. Avery, a Canadian American Physician and medical researcher proved that DNA was the key to biological transformations. From this news, Chargaff realized that the DNA in a cell could be the major part in the concept of genes and hereditary information.

Before Chargaff's experimentations and conclusion on DNA, there were only two main facts known about DNA:

I: DNA is contained in the nucleus of the cell

II: In addition to Sugar-- 2 deoxyribose and Phosphate, DNA is composed of two bases: pyrimidines (Cytosine, Thymine) and purines (Adenine, Guanine)

Other information included was only about important experimental methods involving paper chromatography and ultraviolet light absorption, which was only recently developed. The specific concept that Chargaff discovered was that Adenine and Thymine exist in equal proportions, and therefore pair with each other (and the same concept applies to Cytosine and Guanine). He noticed that the bases' are expressed with the fact that purines are equal to pyrimidines, meaning that Adenine with Thymine are equal to Cytosine and Guanine.

Chargaff also made the conclusion that DNA in the nucleus of the cell is what carries genetic information instead of proteins. He also stated that there must be many different types of DNA molecules in different species. His arguments for this were that although there are only four different nucleotides instead of twenty proteins, the number of different proportions which could exist and many different orders in which it could represent the DNA strand provides a basis of complexity sufficient for the formation of genes.

Chargaff's conclusion revolutionized biological sciences, resulting in the important discovery of the structure of DNA, made by Watson and Crick. Watson and Crick said that due to the fact that Adenine and Thymine always exist in the same proportion, they must always be bonded together, which is also similar to Cytosine and Guanine. This statement, which was established by Chargaff's discoveries, contributed to the discovery of the double helix structure.

Although Chargaff's main interests were about the living cell and personally considering himself as a natural philosopher, Chargaff did many research projects in areas of biochemistry. Besides DNA, he worked on lipids: molecules that form fats, particularly studying on the role of lipid protein complexes in metabolisms, and also did work with the thromboplastic proteins (enzymes that initiates blood coagulation).

Chargaff received many honorary degrees for his contributions to biology, including from the University of Columbia and the University of Basel in 1976. He became a member of many scientific societies including the National Academy of Science, and won many awards, including the Pasteur medal in 1949, the Charles Leopold Mayer Prize in 1963, and the Gregor Mendel Medal in 1973.

Chargaff was considered a rival of Watson and Crick, although he personally believed in no rivalry towards each other, but rather as colleagues in similar fields. This idea of rivalries became popular as when Watson and Crick won the Nobel Prize but he did not.

Chargaff died in June 2002.

So what is DNA?

Deoxyribonucleic Acid

As said before, polynucleotides formed the nucleic acids DNA and RNA. DNA stands for Deoxyribonucleic Acid.

(DNA) **Deoxy*rib*onucleic Acid**

Deoxy: Without (Deo) oxygen (xy)

Ribo: with Ribosomes

Nucleic Acid: Contains Nucleic Acid

DNA is the essential of heredity. More specifically, DNA carries genetic information. Genetic information that DNA carries is what determines traits like the color of your hair or shape of your face. Meaning, DNA is basically the substance that contains the information that makes organisms look different from each other. In organisms known as eukaryotes-- organisms in which the cell(s) with genetic information is DNA in the form of chromosomes, DNA may be found in a place in a cell called the nucleus. Due to the cell being very small, and due to there being many pieces of DNA in one nucleus, each DNA molecule fights tightly within the nucleus of the cell. This form of DNA-- the fact that it is tightly packaged, is called a chromosome.

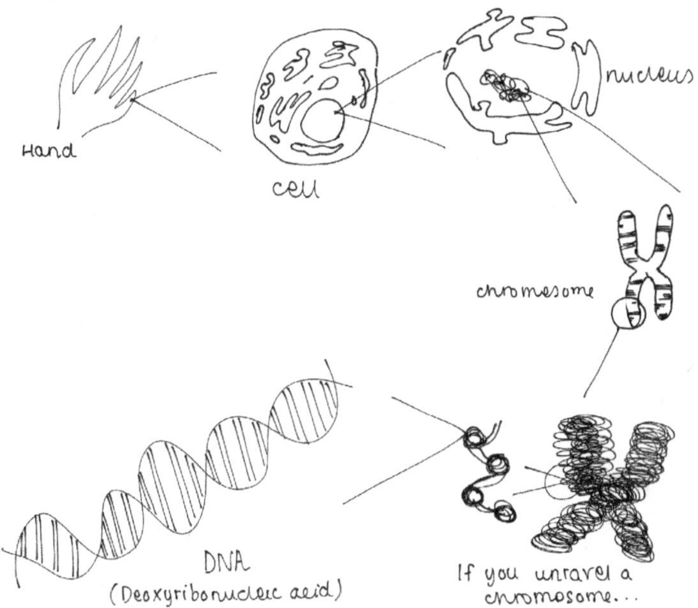

Question: If Genes, Chromosomes, and DNA are basically the same thing, what is the difference?

The reason for all of these different names is because of the history of how these forms were discovered.

People were searching at first for whatever transmitted heredity-- information about what living organisms' descendants will inherit. That is the concept of genes. But what was really discovered were Chromosomes. When examining a cell, the cell always appears transparent, so, scientists created the solution of injecting dye in the cells. This will enable different parts to be visible. The first scientists performed this experiment, they found that a substance within the cell took a very deep color, more deeper than other parts of the cell. This was the chromosome.

Chromosome

Chromo: color

The man who discovered chromosomes was Walther Flemming.

Walther Flemming and the Discovery of Chromosomes

Part I

Walther Flemming was a German anatomist who is known for his contribution to the research on mitosis (when cells create copies of each other), during the late 19th centuries. He was born in April 1843, in Sachensberg/Mecklenburg, Germany. His father was the well-known Carl Friederich Flemming, a famous psychiatrist and neurologist.

Flemming enjoyed studying literature and philosophy, however decided to study medicine due to his family values. He attended the University of Göttingen, Tubingen, and did clinic training in Rostock. While studying in Rostock, Flemming studied historiological and zoological preparation, under the supervision and guidance of Franz Euhard, a professor who was influenced by Maz Schulz (who was known as "one of the first cell biologists"). During this, Flemming learned about constructive criticism, avoidance of speculation, and how to carefully observe during experiments. Flemming was also influenced by other professors and scientists, including Gustav Schwalbe, who implanted in Flemming the idea that cells were the fundamental unit of life (although that was not known yet).

During short periods, Flemming assisted in teaching anatomy and histology in Wurzburg and Amsterdam. In 1870, he was offered the position of leader of dissections and anatomical preparations. By the end of that year, he became an official academic teacher at University. Besides his famous styles of teachings, Flemming was popular due to his talent for art. Students were known for saying that Flemming "brought cells, organs, organisms to life on the blackboard" (Paweletz n.d.). Flemming's drawings were published with his essays, and books.

In 1872, Flemming traveled to the Germany University of Prague, in which he presented histological lectures, seminars, and courses. There, Flemming started to make detailed investigations into cell division, known as mitosis (spoken specifically about in the next section). However, since the end of the German Revolution in 1848, nationalism started growing in the Czech republic. Students of the university Flemming worked to create a university called the Czech University in Prague. Due to this, many German professors preferred to return to Germany. Flemming, however, refused. As he had hoped, he was recruited to be a vacant chair of anatomy at Kiel (1876). The University of Kiel was a small university at the time, which lacked in microscopes, scalpels, and other scientific, good quality tools and equipment, giving Flemming a struggle to work in a hard environment.

In Flemming's late forties, Flemming developed a neurological disease which he never recovered from. At the turn of the century, Flemming retired from his teaching career, as his condition got worse. In August 1905, Walther Flemming died at age 62 in Kiel.

By the time of Flemming's death, Flemming's institute that he worked at became the leading centre for research into histology, cytology, comparative anatomy, and through Flemming's discoveries-- mitosis.

During Flemming's time learning about cell division, Flemming was able to discover about spindle fibers, chromosomes, and "regeneration of tissues and organs that occur by cell division"

Part II

When Flemming began his research into cell division, Biology was just beginning to popularize throughout the science world. In the beginning of Flemming's career (1868), Flemming was mainly interested in the sensory of molluscs (or snails). During that time, he also studied adipose tissue and stated its 'character' as connective tissue: "tissue that supports, protects, and gives structure to other tissues and organs in the body." Flemming also analyzed how lupid droplets are products of cellular metabolism. When Dlemming's focus for his research was on behavior of individual cells, the research on the process of cell decision started to come to light. In 1873, Schneider (another biologist), sketched important steps on cell division, in which he drew nucleus transformations, which are assembled at the center of the cell (inside is the DNA). Flemming saw this, and described this more in detail. Flemming said that Schneider's sketch showed that there was a network within the nucleus, which transforms into two groups-- resulting in two nuclei appearing in one cell (and then when it breaks off, you get two cells).

By studying wounds and scars, Flemming with the help of his students found accumulation of dividing cells in tissues and concluded that the regeneration of tissues and organs occurred by cell division.

--

Notes: DNA was not known at the time. However, they were already onto it, knowing that cells held genetic information, as they knew which organs to form.

--

However, due to the lack of monographs on histological methods not being published, Flemming had to design methods to facilitate his observations. Flemming experimented with many acids to find the appropriate fixative for preserving structure that he had seen in living cells. He used a mixture of chromic acid and glacial acetic acids. This mixture is now known as the "Flemming's solution".

In 1878 and 1879, Flemming published two very important papers, in which in the second paper, Flemming mentioned the wording "indirect nuclear division" (He noted this due to observing the transformation of nuclear content, which had to take place before fusion could occur-- Meaning, the split between nucleus and cytoplasm (the soupy goopy stuff in a cell)-- before this was generally assumed, however, Flemming was the first to acknowledge it as indirect nuclear division). The methods that Flemming developed allowed him to recognize a temporary stance on the nucleus, as it can easily be stained. He called this stainable material *Chromatin* ("stainable material"). As for the structures that remained unstained, Flemming called *Achromatin* ("sustainable material").

From this discovery, Flemming published his famous book: *Cell Substance, Nucleus, and Cell Division.* The impact on this book was major, creating a fundamental foundation of further research into mitosis. Flemming called the cell division process (which was specifically observed alterations within nucleus), Karyomitosis, which means "threadlike metamorphosis". Flemming called the arrangements of nuclear threads *mitosen.*

In 1888, Heinrich Wilhelm Gottfried von Waldeyer-Hartz (a German anatomist) renamed the name of mitosen officially to the current world-renowned scientific name-- *Chromosome.*

Heinrich Wilhelm Gottfried von Waldeyer-Hartz is the 'official' man who named Chromosomes.

Later, it was discovered that Chromosomes are acidic. This acid in the chromosomes was named Deoxyribonucleic Acid or DNA. Watson and Crick were the people that determined the fact that DNA holds genetic information (explained more in Section 2).

Later, it was discovered that inside any DNA molecule, there are some parts that actually don't transmit genetic information, while others do. The parts that do carry hereditary information are known as genes.

In every living creature, the function of DNA is to produce proteins. "We can't explain what makes a human a human, or a frog a frog, from individual genes themselves".

Genomes are the entire set of genetic information within one living organism, which gets passed from one generation to the next.

The genetic information is stored in the bases inside the double helix structure (explained more in Section 2). This genetic information produces proteins, which are the substances that differentiate humans from other humans and animals.

Peter: So, now we are finished with the first part of this book. Getting good, do you think Boys? Tink?

Tinkerbell and the Lost Boys were already fast asleep. Clearly, they were still just children who were not interested in what is inside their bodies, but are rather interested in what they are physically able to do-- eat, fight and sleep.

Peter Pan went to his bed. He was now fascinated by what the grownups from where Wendy lived were doing for fun. He continued reading on his own, still stuttering on some complicated words, but moving on, imagining how Wendy's world is as it was.

Section 2: Proteins, Cell Division, and Protein Synthesis

Peter, now fingering the words he was reading, continued in his head, trying to make no sounds as the Lost Boys were sound asleep.

Protein

Proteins are produced when genetic information in DNA is synthesized and "put to work". They are big molecules that are made up of twenty different types of amino acids (the number and sequence of amino acids linked together form proteins and determine the type of protein). There are thousands of different proteins in a single human cell. Amino acids are twenty different types of side chains. All of the amino acids that are not side chain amino acids, all others are identical. There are two basic types of side chains: Hydrophilic and Hydrophobic.

Hydrophilic	**Hydrophobic**
"Water-loving"	"Water-hating"
(Dissolves in water)	(Does not dissolve in water)

Amino acids are linked together by peptide bonds. The differences in sequence of the linked amino acids determine the properties of different proteins.

All cells are all fully filled with water, which means that hydrophobic amino acids naturally fold in towards the center of the proteins, where there is less water, whilst hydrophilic amino acids remain on the outside of the cell, in contact with water. Different proteins determine living organism's traits. Meaning, the differences between proteins are what create different eye color, hair color, etc.

The protein in DNA is identical for every single living being on the Earth! The difference, which is why animals are all different, is because of different proteins produced for different organisms. Different proteins are produced from different sequences from DNA's base sequences.

Cell division

Cell division is the process of how genetic information is passed on from different cells. DNA has to reproduce itself, and create 'offspring' with old genetic information. Cell division does not happen randomly. There is a special schedule in which the cells grow or divide.

If DNA would split into cells, the new cell would only have half of the original genetic information of the parent cell. The only way DNA can pass on its genetic information to newly formed cells is by duplicating its information to the new cell. There are four main stage to the cell's cycle or cell division:

I: A new and fresh cell produces protein (*The Production of Proteins is called Protein Synthesis*)
II: The DNA of that cell copies itself, forming two individual DNA chains whilst the cell keeps on making protein (*This process is called Replication*)
III: The cell grows bigger whilst the two DNA chains come further apart from each other. The cell continues to produce proteins
IV: The cell splits into two new cells, each with the same DNA. The proteins production stops due to DNA being too tight to read the genetic information

There are two important steps that happen during cell division: Replication and Protein Synthesis (explained in Section 3). Replication is the process of making a perfect copy of the DNA base sequences to create two identical cells with each having an identical strand of DNA. This process preserves and transmits genetic information. Protein Synthesis reads the DNA case sequences and uses it to produce corresponding proteins, putting the genetic information to use. Protein synthesis is what makes everything and everyone differentiate from each other.

Protein Synthesis

Proteins are produced when genetic information in DNA is read and put into work. Proteins are big molecules made up of twenty different types of amino acids. The number and sequence of the amino acids are linked together to form a protein and determine what type of protein it will be. There are thousands of different types of proteins in a single human cell. The different types of proteins are essentially what differentiate humans from animals, or plants from insects: differences in eye color, hair color, body shapes, and so on.
In other words, genetic information is transmitted through proteins.

There are two scientists that have been credited for the discovery of Proteins. One of these scientists was Gerardus Johannes Mulder, a chemist from the Netherlands. He studied medicine at the University of Utrecht during the early 1800s, in which he wrote a dissertation about the action of alkaloids-- a certain class of organic compounds. He practiced medicine in Amsterdam as well as Rotterdam, in which not only did he perform medical experiments, but also taught botany to students. In 1828, Mulder became a lecturer in botany, chemistry, mathematics, and pharmacy. In 1840, Mulder became the professor of chemistry at the University of Utrecht. He retired in 1868, and passed away in 1880. Throughout his career, Mulder discovered and researched the idea of proteins. He found the proteins to be very big molecules-- bigger than what chemists thought before. Through experimenting on the composition of some animal substances, he noticed the chemical composition of several proteins, and understood that all of them have the same composition. Mulder concluded that all proteins consist of one primary substance, and that animals get most of their proteins from plants. He also learned about components of proteins and smaller molecules. He found out that these molecules had amino and carboxylic acid functions. These types of proteins are now known as amino acids. The other scientist credited for the discovery of proteins is Jons Jacob Berzelius. He suggested to Mulder a name to call what Mulder discovered: "proteins". This was due to the question that if proteins were a fundamental substance, then the word "primary", meaning"in the lead", or "standing in front" from the Greek word "proteios" is most appropriate.

The way DNA's genetic information gets put into use is through the production of proteins. About 70% of your body consists of water and about 18% of your body is proteins. Although there are so many different types of proteins in your body, the basis by which the genetic information of DNA is produced is identical to every single living organism on Earth! This means that the process of the creation of proteins from DNA are the same for all organisms, but the form that protein makes the organism form into are all completely different. This is because different proteins are producing different types of organisms. The differences are in the base sequences inside DNA (the A-T-G-C bases). Each type of organism has a different "lineup of genetic information". Not only that, but proteins are by which genetic information creates individual organisms.

As stated before, the two important steps in cell division, or the duplication of more identical cells are replication and protein synthesis. Protein synthesis reads the DNA bases sequences (the parts inside the double helix: A-T-G-C) and uses it to produce corresponding proteins for a specific body part, putting the genetic information to work.

Question: So then why do we look like our parents or siblings?

Your cells contain genetic information from your parents-- from your biological mother and father. Some of their genetic information will be transmitted into your genetic information, so when protein synthesis occurs, you may have similar traits with your parents or siblings!

So, how the reading genetic information of DNA works is that the DNA makes proteins based on the genetic information and then it transmits that information by copying the DNA and dividing it into many cells.

Watson and Crick's discovery of the complementary base pairs A-T and C-G did not only help scientists understand the structure of DNA, but also by what means genetic information gets transmitted.

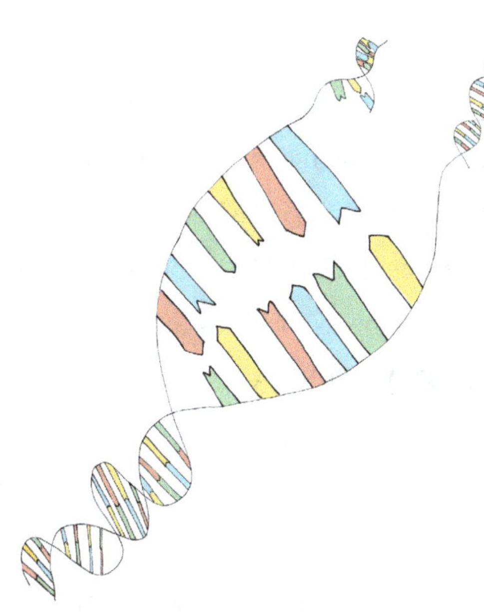

Basically, the complementary relationship between the two strands of the double helix structure is what makes DNA possible to replicate itself. When DNA goes under replication, the strands separate and a new complementary DNA strand is synthesized on each of the two original strands from the original double helix structure. Each original strand is kind of like a mold or template for the newly created future DNA, which can be produced precisely because of the base pairs.

Protein Synthesis

After Watson's and Crick's impactful discoveries, scientists began wondering, how exactly is genetic information read? Molecular biologists began studying the relationship between DNA and protein, and it was first thought that genetic information of DNA must be the information that synthesizes proteins. That means that various kinds of proteins must be synthesized according to the sequence of the four bases inside DNA. Scientists quickly understood that this theory was incorrect because it was already known that protein synthesis does not take place in the nucleus of the cell where the DNA is located. Protein synthesis happens in the cytoplasm-- what surrounds the nucleus inside of the cell. Biologists experimented more and more. They knew that it had to be some type of substance that receives genetic information from DNA and carries it out to the cytoplasm where the proteins are synthesized.

Scientists decided on RNA (Ribonucleic Acid)

RNA is a nucleic acid that is known to be one of the main substances inside the nuclear membrane. Other substances include DNA and protein. RNA also consists of a chain of nucleotides that consist of a base, sugar and phosphate, just like DNA. Not only that, but there are also four different types of bases inside of RNA, the same as the ones in DNA, except for one. In fact, there are three main differences between DNA and RNA:

I: DNA's structure is the double helix, whilst RNA's structure is just a single strand

II: Instead of one of the bases Thymine (all others in RNA are the same as the ones in DNA), RNA has uracil (U).

III: The sugar in RNA has one more oxygen atom than the sugar in DNA (RNA stands for Ribonucleic Acid, whilst DNA stands for Deoxyribonucleic Acid. This means that DNA does not consist of oxygen, while RNA does)

Basically, DNA is RNA that lacks one oxygen atom.

About half a year after discovering the double helix structure and the Complementary bases, Francis Crick made a hypothesis which is known as the 'Central dogma'. His hypothesis was that DNA's information is transcribed into RNA, which is conveyed later as protein:

$$DNA \rightarrow RNA \rightarrow Protein$$

At first, scientists thought that RNA might form a template for protein synthesis by folding in on itself and forming indentations that fit twenty different amino acids. But what Crick believed was that his method would not allow for clear distinctions to be made between different amino acids, similar to some chemical structures. He hypothesized that there must be some sort of "adapter molecule", which picks up amino acids before they are incorporated in the protein chains.

$$DNA \rightarrow RNA \rightarrow Protein$$

Adapter carrying

amino acids?

After many experiments, scientists found a small RNA molecule in which the amino acids would stick to before they became a part of the proteins. Since it was a small RNA that transfers amino acids, it was decided to be called tRNA. Basically, Crick's description completely described tRNA from his hypothesis.

$$DNA \rightarrow RNA_{tRNA} \rightarrow Protein$$

After the discovery of tRNA, scientists discovered a new particle called Ribosomes, which were now known to be able to conduct the protein synthesis. It was discovered that these ribosomes contain RNA, so it was dubbed as ribosomal RNA or tRNA. Since ribosomes perform protein synthesis, scientists at first thought that rRNA in ribosomes might be what determines the sequences of amino acids that go into the protein. However, this was proved to be incorrect due to ribosomes all being very similar, so they would not be able to account for differences amongst proteins of different organisms.

DNA $\rightarrow$ RNA ᵗᴿᴺᴬ $\rightarrow$ Protein

rRNA-- ribosome factories

of protein synthesis

In 1960, Biologists found a third type of RNA cells that had been infected by T4 phage (escherichia virus). A phage is a virus that infects bacteria (EX: E.Coli-- a unicellular bacteria that is very commonly used within experiments). When phage infects the E.Coli cell, it uses the ribosomes and tRNA of the E.Coli to make another RNA-- T4 RNA inside of the bacterium. Scientists figured that it must be T4 RNA that gets the genetic information from NA and acts as a template for protein synthesis. Later, it was realized that the RNA that played the same role in cells is not infected by phages, and that it was the RNA that carries DNA information. This type of RNA became known as mRNA (messenger RNA). Of all the RNAs there are, 85% of all RNA is rRNA, 10% of all is tRNA, and 4% is mRNA (which was why it took a long time to discover).

In total, there are three types of RNA, which all play a different role. And it was then determined that the genetic information in DNA was first transferred to mRNA and then to protein.

DNA $\rightarrow$ mRNA $\rightarrow$ Protein

mRNA= carries DNA

tRNA= carries amino acids info

rRNA= forms ribosome factories

of protein synthesis

Consider. . . There are twenty amino acids but only four bases, so how do they specify all of the different amino acids?

If you pick 3 bases out of the four at a time to represent one single amino acid-- 4*4*4=64, which is more than enough combinations to represent twenty amino acids (Two bases would not work though because there are not enough options: 4*4=16 variations). In 1954, molecular biologists had already hypothesized that it took a sequence of three bases to specify a particular amino acid. SO the next question was-- Which sequences correspond to which amino acids? To this question, scientists performed a revolutionary experiment. In 1961, scientists created artificial RNA sequences, which consisted only from the uracil bases (U-U-U-U etc). They were doing this to attempt to get the ribosomes to make a protein with it. As a result, the polypeptide chain made up of two of the twenty amino acids created phenylalanine. Then, scientists began testing another artificial RNA sequence after another. By 1966, scientists figured out which amino acid corresponds to each of the 3 base sequences. These 3 base units are called codons.

Scientists also determined which codons determine where the process of replication starts and stops. Eventually, they succeeded in deciphering the genetic code for 64 possible codons-- all options.

Biologists discovered that the genetic code can be applied to all organisms. Meaning, that the process of replication, transcription and protein synthesis is all the same process. This led molecular biologists to conclude all living organisms came from a common ancestor-- the first original cell.

In 1970, scientists discovered a certain type of enzyme now called DNA restriction enzymes. This type of enzymes are capable of cutting the DNA molecule at a specific point in the sequences of the nucleotides. This discovery 'gave birth' to a new field of biology: recombinant DNA technology, which allows for scientists to analyze and manipulate DNA of large, molecular organisms (EX: humans). Specifically, not just a simple DNA organism like E.Coli. This also developed a method of determining sequences of DNA bases, known as sequencing. From these new abilities in the biology field, working with Nucleotides and Nucleic acid molecules became much easier to use.

Another discovery made was the discovery of Eukaryotic cells, cells with a nucleus, like plant or animal cells, genes containing some sections that are used for protein synthesis while other sections were not used. The sections that were used are called Exons and the not used ones-- introns. After the cell transcribes a gene sequence into the RNA, RNA goes under a process of splicing, which is when unneeded intron sequences are cut out of the RNA molecule. Prokaryotic cells-- cells without a nucleus (EX: E.Coli) usually do not have introns, and thus splicing does not occur for protein synthesis.

Protein Synthesis

There are four steps in protein synthesis.

Step 1: Transcription   Transcription of DNA information into RNA

- The first double helix of the DNA in the cell nucleus opens and the base sequence contains genetic information (which is copied in the RNA molecule). This process is called Transcription, which is facilitated by the RNA polymerase. The RNA polymerase is an enzyme that acts as a catalyst for the process

CHART: showed which RNA base corresponds with which DNA bases

DNA: A   T   G   C

RNA: U   A   C   G

Notice: instead of Thymine for RNA bases-- Uracil (U)

In DNA, the entire molecule replication is copied. But during transcription, only some sections of DNA that actually function as genes are copied.

Step 2: RNA Splicing

In Eukaryotic cells, genes come in two parts, in Exons and in Introns. Exons carry the information for protein synthesis, whilst introns do not. After DNA's genes have been transactive to the RNA molecules, introns, which are not needed for protein synthesis, are removed from the RNA, leaving only the Exons. This process is called RNA splicing. After RNA leaves after splicing, the mRNA moves outside of the cell nucleus into the cytoplasm, which serves as a mold for protein synthesis when it occurs.

Step 3: Translation of mRNA information into the Amino Acids

The process known as translation, which is the process of when the information inside the mRNA is converted into the sequence of amino acids gets broken down into four steps:

- **I: The mRNA attaches to a ribosome.** The sliced mRNA then moves outside of the nucleus and attaches itself to the ribosome located in cytoplasm. The ribosome-protein synthesis factory is in the form of a large clump of rRNA and protein (this is made out of two subunits: 1 large and 1 small). The ribosome manufactures protein by translating information carried by mRNA into a sequence of amino acids, which will make up the protein molecule

- **II: The ribosome translated genetic information carried by the mRNA.** The ribosome reads the base sequence of mRNA and three bases, and translates each codon into one amino acid. Scientists discovered that the resulting code allows for several codons to correspond to the same amino acid.

- **III: tRNA carries amino acids to the ribosome.** Before the ribosome translates the DNA information from the mRNA base sequence into amino acids, the tRNA carries the amino acids to the ribosome factory and lines up into place. Aminoacyl-tRNA synthetase, is an enzyme that couples amino acids to one end of a tRNA molecule. This enzyme 'knows' which of the 20 amino acids to attach to which tRNA. On the other end though, the tRNA carries anticodons-- a triplet of bases that are complementary to codon on mRNA. The tRNA is bound to the appropriate amino acids by aminoacyl-tRNA synthetase. This thus creates a one on one correspondence between codons and amino acids.

- **IV: Ribosome links amino acids together.** So, everything-- the mRNA, tRNA and ribosomes are now in place. Now, the tRNA molecule copied to the one amino acid on the other end (which has a codon) corresponds to one codon on mRNA molecules. The tRNA molecule arrives at the ribosome and binds to corresponding a codon on to the mRNA. The base sequence information represented by one codon on mRNA is hooked to the corresponding amino acids. Then, another tRNA molecule corresponding to the next 3-base codon on the mRNA arrives with another amino acid, which adds on to the long chain. The way these connect into a chain is by having two adjacent amino acids linked together by a chemical reaction in the ribosome. This process repeats over and over until the ribosome has produced a long amino acid chain, known as polypeptides.

Step 4: Completion of Protein Synthesis. Ribosomes continue the process until there is a codon called a stop codon on the mRNA molecule that indicates that the translation strops there. The mRNA, Ribosome and polypeptide detach from each other and go their separate way. The polypeptide chain twists into itself in a 3-dimensional shape inside of the cell. This is a natural tendency because some of the amino acids that make up polypeptides are hydrophilic (water-soluble) and the others are hydrophobic (water-insoluble). The cells are full of water, so the hydrophobic amino acids fold in towards the center of the polypeptides (as if barricading itself) and the hydrophilic amino acids remain outside, having contact with the water. This causes the polypeptide to fold into a 3-dimensional shape, resulting in the creation of proteins!

This cycle is ongoing, and occurs all of the time. Genetic information transcribed to RNA and then to protein. Proteins help with the replication process of DNA.

Question: So, how did the original first cell get started?

Peter Pan, now thoroughly entertained, thinking about how those scientists came up with these strange ideas of how his body ticks, noticed that the Lost Boys were waking up as the sun started coming up. He flew out of his room and sat at the top of a tree with Wendy's book, eager to learn about the first creatures on Earth.

Section 3: Cells and Organisms

The first cell

As now known, the ongoing cycle between DNA, RNA and proteins go on forever, in which genetic information transcribed from DNA to RNA, and then to proteins, and proteins help with the replication process of DNA. For genetic information to be able to work, the cells need substances that function like glue: enzymes (EX: DNA Polymerase and RNA polymerase). Without enzymes, DNA cannot get replicated and proteins cannot get synthesized (Note: those enzymes are made out of protein). This means that although DNA may hold genetic information, however without enzymes, the genetic information is useless as it cannot be passed on. However, enzymes were not the first creation that started the first cells. Enzymes are proteins and DNA holds the instructions for the creation/synthesis of proteins with amino acids. In conclusion, scientists still don't know how the first came into existence yet. It is also so far unknown about what was the first living organism or how the first cells formed, but there is a theory that is quite reliable.

Cells today are made out of amino acids, nucleotides, etc. These substances are known as organic matter, which are made out of organic compounds. Organic compounds contain carbon, so all living organisms are made of carbon compounds. This means that organic matter probably already existed before the first cell was created, and scientists have found proof of that. In 1955, a scientist named S.L Miller performed an experiment in which was a simulation of the environment of the 'ancient' Earth. This experiment concluded/suggested that the ingredients for life had existed on Earth before life appeared on Earth. This conclusion is just a theory. However, scientists accept this as a fact.

S. L. Miller (or Stanley Lloyd Miller), was born in California. Miller was always interested in the origins of the Earth and how chemistry played a role in it. He went to the University of Chicago in 1951, and studied under the professor of chemistry. While writing his thesis, he became more interested in understanding how chemistry helped biological cells work many thousands of years ago. In 1953, Miller performed an experiment in which he tried to understand what occurred in the primitive Earth billions of years ago. What he did was he "sent an electrical charge through a flask of a chemical solution of methane, ammonia, hydrogen and water" (PBS). This experiment resulted in the creation of organic compounds and amino acids.

So, organic matter existed before the existence of life. This is known as the prebiotic synthesis. Organic amino acids and nucleotides were created from the lightning, volcanoes and ultraviolet sun rays. They began joining together with molecules to form bigger ones, known as Polymers. Nucleotides have a natural tendency to bond with each other, however nucleotides do not. However, scientists believed that the harsh environment during the times of pre-living organisms as well as the molecules would be subjected to hear or would encounter substances called catalysts, which promote chemical reactions linking molecules together into the polymer long chains (referring to chains of amino acids as proteins and as nucleotides and nucleic acids like DNA and RNA).

A living organism is only classified as a living organism as long as it can replicate and must be able to evolve. It seemed unlikely that the first life forms had a perfect system or replication, so there would be many errors when replication would occur. Due to these errors, organisms would tend to produce slightly different organisms when they replicated. The offspring that dominated the environment would survive. This is known as evolution. From these theories, scientists decided on the fact that the first organism had to be capable of self replication.

For molecules to replicate themselves, they must have abilities to catalyze reactions which will result directly as well as indirectly to their own replication. Without catalysts, there would be no chemical reactions. Catalysts are substances that speed up a rate of chemical reactions without undergoing change themselves. Nearly all reactions that occur inside of organisms depend on some type of catalyst. For molecules of the first living organisms to be able to produce another group of molecules like themselves, they "had to be able to act as a catalyst for their own replication reaction". This type of group of molecules is called the

Autocatalytic system. Autocatalytic systems have two main properties:

I: A system that can catalyze itself

II: A system that can decay if it cannot perform its own replication, due to running out of materials or suffering from sudden changes in the temperature of environment

Autocalytic systems are very life-like and are associated with the qualities of life.

The most common type of catalysts in cells currently are proteins. Proteins can take the form of many shapes in 3D-- this property makes them very effective catalysts. Proteins however, cannot make more proteins all by themselves. Proteins are effective catalysts but cannot qualify as the first living organism on Earth.

As we now know, DNA has the ability to self replicate, however it cannot perform catalysis. RNA can produce complementary base pairs, and can self replicate (it has nearly the same structure as DNA). As for if RNA can perform catalysis-- it was thought of not having that ability, due to having such a similar structure as DNA, which cannot perform it.

Self splicing rRna of the tetrahymena

In 1981, scientists observed an rRNA molecule in a Tetrahymena (a critter that is one cell, covered in hair, with a round cytostome (cell mouth) and is about 15-70mm by 6-50mm l). It was observed that rRNA uses a part of itself as a catalyst to cut down its own introns and splice itself. This became known as "self splicing", which was thought of by Thomas Cech and his work colleagues. So, it was discovered that rRNA was acting like a catalyst. This made RNA hypothesized as the first living autocatalytic system.

However, RNA is not the same as other catalysts . Tetrahumena's RNA is different from other catalysts because it cuts itself off after the stop codon. Real catalysts influence chemical reactions but do not change themselves. The intron segment is a part that gets cut off by self-splicing and is not used by the RNA. This makes RNA act like a fully fledged catalyst. The intron's role is to lengthen and/or shorten small nucleotide chains that are called oligonucleotides. Introns don't change at all. And this determined the fact that introns are actually the real catalysts. In conclusion, scientists knew that RNA stores information the same way as DNA and it folds in on itself into a three dimensional form, allowing it to perform catalytic functions. But, in other cells, introns are the catalysts.

Self replicating molecules undergo natural selection

RNA molecules are capable of creating complementary base pairs within themselves and can fold into 3D shapes. But it is not that plain and easy. Some folded shapes tend to break more easily than others, and errors can also occur during the copying process. These changes cause RNA to produce versions with itself in different shapes, producing interesting results. In the early 1990s, scientist conducted the following experiment:

- **Step 1:** Prepare a column that is packed with beads of resin. Scientists performing this experiment had a specific test substance that stuck to the resin
  Resin column is a pipe that is packed with resin beads
- **Step 2:** Artificial RNA molecules that are created with random base sequences get passed through a column, which leaves behind only the RNA that bonds to the test substance
- **Step 3:** Scientists then would extract RNA molecules that remained in the column and used them as template to replicate and produce more RNA molecules
- **Step 4:** Newly replicated RNA gets passed through the column again
- **Step 5:** By repeating this process over and over again, scientists accumulated a large amount of RNA sequences that would bind to their certain test substance

When RNA gets repeatedly replicated and passed through the first column, the amount of RNA increases. The theory for this reason is due to the shape of RNA. The conclusion of this experiment said that RNA contains information in the form of base sequences that can pass through replication (informational reason) and that RNA has a specifically folded structure that allows it to interact with the surrounding environment selectively (functional reason). These are RNA's two properties and both of these reasons are necessary for evolution to occur in the living organisms.

Scientists performed another experiment in which they put phospholipids and water into a test tube, and found out that phospholipids are gathered together to form a bilayer membrane. This suggested that the first cell membranes formed from similar phospholipids in the primordial soup. Cell membranes are made out of phospholipids, also known as molecule fat. Phospholipids are made out of hydrophilic and hydrophobic bonds as shown in the diagram:

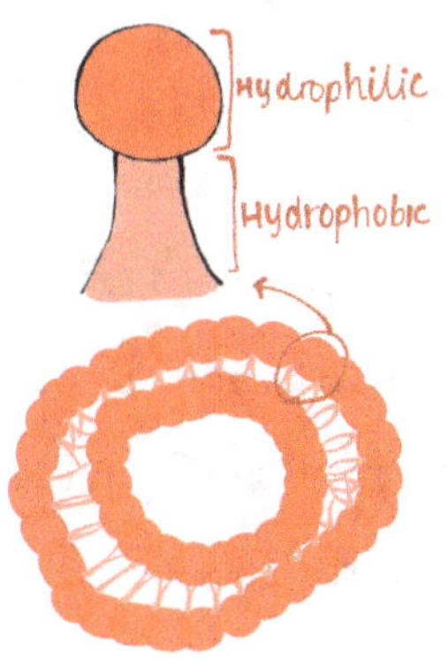

When they are immersed in water or the primordial soup, the hydrophobic ends stick together as closely as possible, whilst the hydrophilic ends face outwards towards the water. The result of the phospholipids is the creation of a natural membrane. Phospholipids have properties that make them formed closed compartments known as vesicles (by arranging themselves in a bilayer, which acts as a membrane) that are isolated from the outside environment. When the first time this process happened, was when the birth of the first cell occurred. Early RNA molecules probably used phospholipid membranes to prevent synthesized proteins from escaping.

The earliest cells were different from today's cells. That was due to the fact that early cells probably had stored genetic information inside RNA molecules, whilst today, DNA stores it.

Overall, DNA is more stronger and stable than RNA due to the fact that DNA is made for double helix structure (double strands) and because it is made of complementary base pairs. This makes it easier to repair the strands that become damaged at some points in time. This occurs when the bases on the other strand serve as a template to restore correct information. During evolution, RNA must have given information-storing functions to DNA, giving it a bigger advantage in survival. And supposedly from that, DNA continues to store gene information today.

RNA → Genetic Information → DNA

Before this was known however, no one found out that RNA actually catalyzes its own replication. Instead what was known is that RNA appears capable of autocatalysis.

It can be presumed that self replicating RNA molecules began collecting somewhere on Earth about 4 billion years ago. Scientists have no doubt that different groups of RNA compete against each other for raw materials that are needed to replicate themselves and produce offspring. To make sure that the molecules that other molecules went up against were "easily defeated", molecules groups would have to replicate quickly and accurately, or produce stable copies of itself that would not degrade instantly. RNA molecules are relatively good at storing information and replicating it but compared to proteins, there is a bigger limit in catalytic abilities. For most organisms, catalysis is performed by proteins. So, there must be some point in which the RNA molecule groups have to incorporate proteins into self-replicating systems. It is not known how RNA found out how to use proteins, but it can be seen that various forms of RNAs worked together. Through this, RNA molecules that developed the ability to direct protein synthesis with superior catalytic abilities had a huge advantage in survival.

Another requirement for the beginning of cells was to have an outer member. Without some form of a membrane, the cell cannot survive as a cell. The reason for this is that certain RNA molecules synthesize proteins for themselves, however, protein has no part in replication of RNA. If competing with others to get the protein first, RNA would need to put proteins into a certain storing place, or enclose it in some sort of membrane. Then RNA would only belong to itself. DNA took over the storage of genetic information from RNA, which is why DNA is still here today, still in a similar form. This is how scientists believe the first cell was formed.

This theory of RNA is just a hypothesis, however scientists are increasingly being convinced that life began with RNA.

Scientists believe that "survival of the fittest" applies to RNA molecules as much as to plants and animals. This is based on Darwin's Theory of Evolution, in which organisms with genetic information that is beneficial to their survival and reproduction are naturally environmentally selected or the organisms survive and those survivors evolve.

The first multicellular organism

It was believed that about 4 billion years ago, the Earth had nearly no oxygen in the atmosphere. Instead, the atmosphere was full of CO2 (Carbon Dioxide), CH4 (Methane), NH3 (Ammonia), and H2 (Hydrogen). At this point in time, the Earth is getting constantly hit by meteorites, rainstorms, lightning, and ultraviolet rays from the sun. However, somehow, through all of this commotion, the first cell was born. "The birthplace of life was the ancient ocean".

The original, first cell created was a unicellular organism, in which the organism is made of only one cell.

So how did the first unicellular organism in the primordial soup evolve into multicellular organisms, to create and evolve into today's living organisms like humans and plants?

The first cells on earth are known as Prokaryotic cells, meaning cells without nuclear membranes. The E.Coli bacteria (spoken about before) which inhabit human's intestines are prokaryotes that resemble very closely the first ever cells, which is why they are used commonly in experiments on unicellular organisms. So how did the E.Coli Bacteria evolve?

Cells had the advantage of being enclosed in a membrane, with exclusive access to produce proteins. Inside the cell, substances causing many different reactions were packed together inside the cell. The cause for an increase in the speed rate of reactions and enabled the cell to divide and reproduce quickly. When the number of cells grew too rapidly, problems started to occur, including the fact that the cells altered the environment that surrounded them.

When the increase of molecules started in the primordial soup, many nutrients and molecules experienced a food shortage or lack of food in the ocean, or at least,at least that was what it seemed to be. Actually, the food had not completely run out. It was just how the bells have to prepare their nutrients before eating it. The cells started making their own food utensils, including enzymes and certain types of proteins. This produces the food that is lacking in their environment. The earliest food preparation process was the earliest form of glycolysis. Glycolysis is the process of breaking down glucose of the enzymes.

The process of food preparation that is conducted by cells is known as metabolism. This process is essentially the same definition of when we eat food. When we eat food, the body digests it in the stomach, then the nutrients get absorbed by the cells, which breaks down the food even more, and in the cells, the nutrients are used as building materials as a source of energy. There are many methods in which cells perform metabolism. For example, since humans don't eat the same food all of the time, their bodies have to be able to get energy from many different nutrients. There are many "routes" that are taken by the metabolic process. This is known as the metabolic pathways.

About 4 billion years ago, the environment around the first cells on earth probably did have everything needed for their survival. Their nutrients required for them to all occur naturally in the primordial soup, but as cells multiplied and started to increase, the cells began to use up all of the available nutrients. The way these ancient cells dealt with the increasing food shortages was by changing themselves to adapt to the environment, which created new types of metabolic pathways. However, now there was a new problem-- "garbage disposal".

When a cell that is enclosed in the membrane produces its own food through metabolic pathways (EX: glycolysis), it produces garbage/waste in forms of H2 (Hydrogen) ions, which are actually protons. If the garbage were allowed to pile up inside of the cell, the cell would become dangerously acidic, which could lead to the killing of the cell.

The solution to dispose of garbage was through the use of membrane proteins. Proteins that are present in the cell membrane utilize energy that is stored by the molecule known as ATP, which is an acronym for Adenosine Triphosphate, to move the protons across the cell membrane to the outside. These proteins that perform this are known as protons pumps.

ATP is one of the most important molecules in an organism. ATP functions as storing energy that can be used inside the cell. ATP is made of the base adenine, three phosphates, and a sugar. Cells are able to use large amounts of ATP, when it is being manufactured by protein or DNA. For cells, "ATP is crucial to survive".

However, a problem occurred with this proton pump method: If a cell would dump out too many protons, the protons would start to flow back to the cell. They would go through the obsolete proton pump. However, cells found a new way that is much better for themselves. Primitive cells could dispose of their waste by the method called electron transport, in which primitive cells could get rid of protons without using ATP as the energy source. Instead of ATP, electrons were used (electrons are produced by metabolic processes as energy sources).

Now cells had found a way to make use of the force with protons flowing from the outside of the cell membrane (So basically: many protons= no protons at all). This is known as proton-motive force. The cells could use the energy of protons to enter the cell through the proton pump to synthesize ATP through a recycling system. In ATP's now new role, the proton pump is essentially the ATP-synthesizing enzyme, known as the ATP synthase.

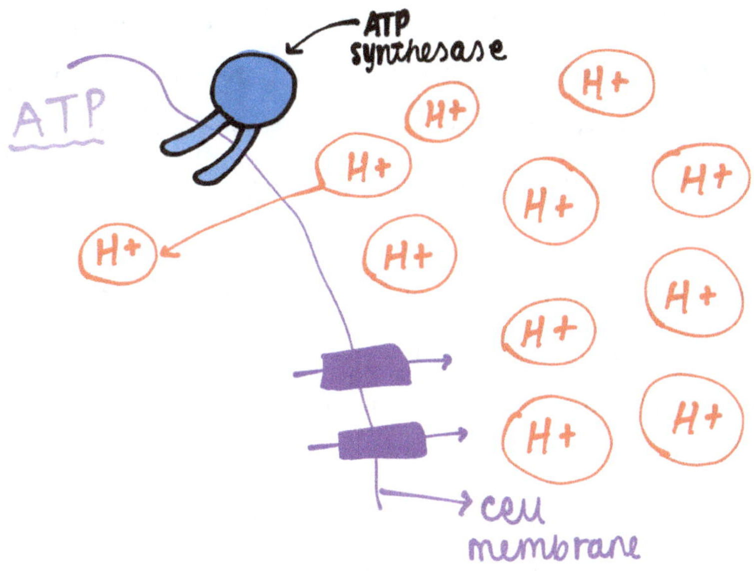

The process of this is that the proton pump would start using ATP's energy to pump protons out of the cell. Then, it would actually create ATP.

The next important point in evolution, occurring about 3.5 billion years ago, was the advent of photosynthesis. As the number of living organisms on Earth increased rapidly, organic materials became very scarce, which was a big trouble for the surviving cells.

As established before, carbon is one of the most important nutrients in all living organisms' cells. As materials, including carbon became very scarce, if the cell could not find a supply of carbon, the cell would die. Main source of carbon was CO2 (carbon dioxide). One way cells would convert CO2 into organic materials was through using the substances NADH and NADPH. These substances were able to produce only a little bit of substances at a time. NADH and NADPH are almost the same molecule (through a chemical perspective). Both substances act as a high-energy electron carrier. As cells need electrons, which were given by the NADH and NADPH to be able to convert CO2 into Carbohydrates or carbs (CH20).

From these abilities, new cells appeared that were capable of utilizing the energy from the sunlight to produce these NADH or NADPH. These cells are known as cyanobacteria. Eventually, cyanobacteria developed the ability to manufacture organic materials from CO2 and NADPH. This was the beginning of photosynthesis, as organisms were now able to produce their own energy even when there is no food for them to eat. Cyanobacteria also produced a by-product of photosynthesis: oxygen (O2).

There was another problem however, that eventually affected all living organisms: the oxygen that was produced by cyanobacteria was poisonous. This causes chemical reactions to occur, and when it reacts with DNA and proteins, it would break down. All oxygen that was discharged by the cyanobacteria slowly altered the Earth's environment. Iron ions in the ancient oceans bonded with the oxygen, forming iron oxide. Big deposits of iron oxide settled to the bottom of the seas.

By about 1.5 billion years ago, all of the ocean's iron ions had been converted to iron oxide. At that point, the oxygen that was produced by cyanobacteria began to accumulate in the atmosphere. When oxygen encountered the ultraviolet rays from the sun, a chemical reaction resulted into the production of Ozone (O3). The ozone is what prevents ultraviolet radiation from reaching the Earth's surface. This ozone that was created is currently known as the Ozone Layer. For every organism on Earth, except for the cyanobacteria, the increase of toxic oxygen in the atmosphere was extremely bad. However, oxygen had a beneficial side effect of blocking the ultraviolet rays. The increase in oxygen was very slow, and the Earth's organisms had plenty of time to develop ways to protect themselves from the toxic effects. As the amount of oxygen increased, cells began to lose the abilities to perform photosynthesis. Yet, organisms were still able to compensate by using the oxygen to break down the organic materials that were produced by photosynthesis bacteria into CO2 and H2O. This resulting energy was used to synthesize ATP. These cells that were capable of doing this are the ancestors of the mitochondria, which now is a very important component in all cells.

When the food that is digested in humans' stomachs enters a cell, it is further broken down by complex processes like glycolysis, and most importantly, the citric acid cycle. When eating, the enzymes inside the digestive system break the food down into smaller molecules (EX:

glucose). These molecules get carried by the bloodstreams to the cells of the body. When inside the cell, the glucose undergoes glycolysis, in which four repetitive steps take place:

- **Step 1:** Glucose converts into pyruvate

- **Step 2:** Enter the mitochondrion

- **Step 3:** Passes through the citric acid cycle

- **Step 4:** Undergoes electron transport

- **Step 5:** Repeat

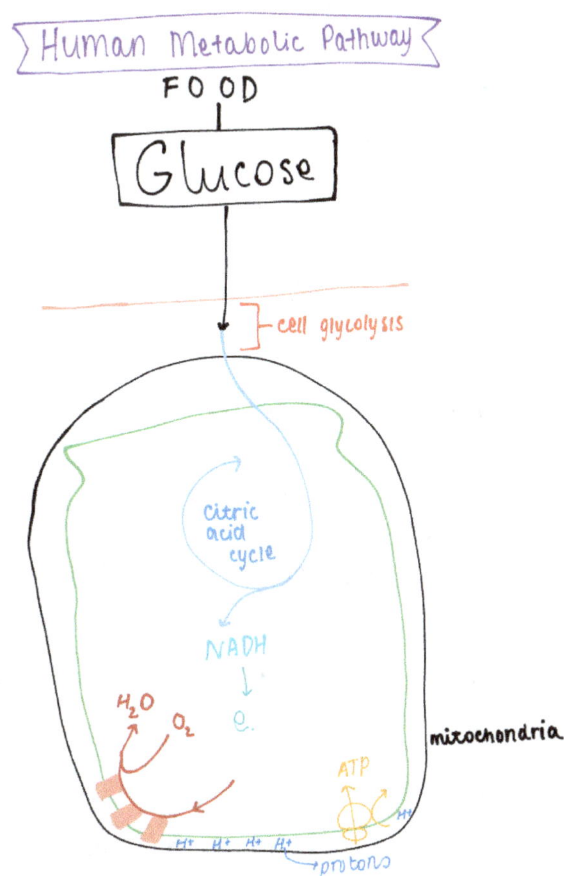

Evolution of Glycolysis

Carbohydrates are made out of polysaccharides, which are basically long chains of sugar. When these are digested, they break down into six carbon sugar glucose that are capable of entering cells. Glycolysis is a 9 part process of chemical reactions that split the six carbon glucose molecules into 2 3-carbon molecules, which are converted into the acid-- pyruvate. This enters a certain type of organelle known as the mitochondrion. Organelles are a "specialized compartment inside the cell". After passing through the next process known as the citric acid cycle.

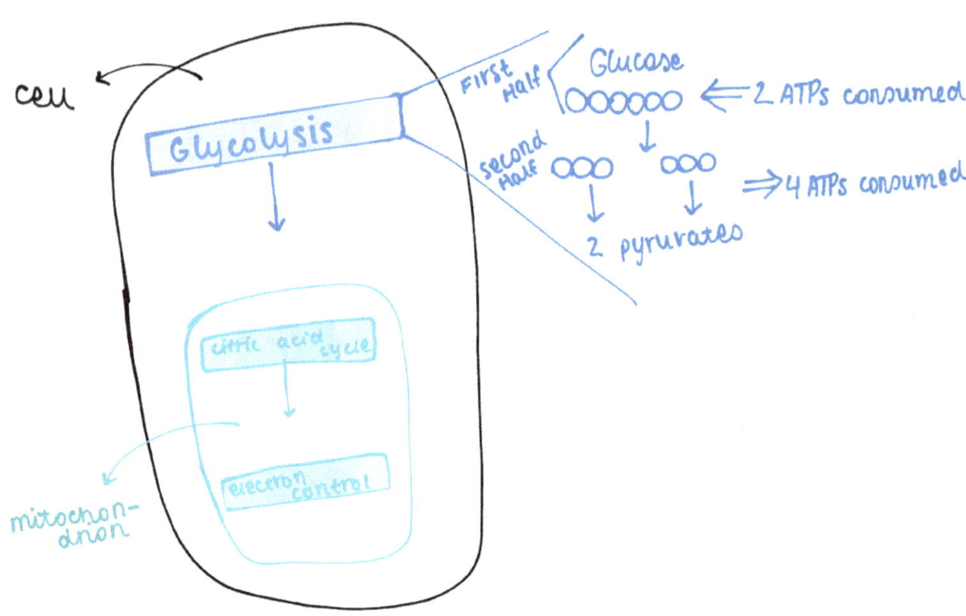

In depth the process of converting glucose to pyruvates:

Part 1: ATP gets consumed

Part 2: ATP is produced

Glycolysis is supposed to be an energy-producing process however it actually consumes ATP.

However, for every two molecules of ATP that is initially consumed, it produces four ATPs (net gain of two ATP molecules).

When looking at the origins of glycolysis, reactions in the second half of the process are the same reactions that were used by the ancient organisms to produce energy. This was done by consuming nutrients that are available around them on ancient Earth and producing just enough energy to keep themselves alive. When the ancient oceans and seas were full of important nutrients, the organisms did not have to get into the trouble of consuming energy in order to produce more nutrients. The primordial soup had all that was needed for the cells to perform energy-producing reactions. This reaction is the second half of glycolysis, which is when the glucose which is now a 3-carbon molecule gets converted into pyruvate. Now that cells had no choice but to figure out the ways to produce energy from the materials that remained, an access to plenty of raw materials was more important to the cell's survival, rather than saving two extra ATPs. As a result, they would consume it, which is part of the first part of glycolysis.

Result of the evolution:

Processes that arose out of the Earth's environment many years ago still survived today in all the cells currently on Earth. Now, current cells use those processes to benefit themselves.

Citric Acid Cycle

This process is one of the most important metabolic pathways in the cell. It is a series of reactions that take place inside the mitochondria of the cell.
The Citric Acid Cycle is not a simple cycle, in which the materials "don't just enter the mitochondrion and get spun around". The mitochondrion is filled with different molecules and enzymes.

In this environment, the first substance produced in the citric acid cycle undergoes a series of seven reactions, with the result of the production of more citric acid. The citric acid cycle also performs two important tasks: it produces lots of energy and it produces materials needed to build the body of the cell.

When there was no oxygen, the citric acid cycle did not exist. Two big developments made the cycle possible:

- **1:** The birth of the enzyme that joined two diverging metabolic pathways into a circle. These two pathways were the ancestors of the citric acid cycle
- **2:** The appearance of huge quantities of oxygen in the atmosphere

The ancient cells had two separate metabolic pathways, which were ancestors of the citric acid cycle. The two pathways were able to produce materials for building the body of the cell, and are able to produce NADH, which serves as a carrier of energy inside the cell. As the ancient atmosphere filled up with oxygen, cells started to evolve, creating NADH to produce ATP. These cells are the ancestors of mitochondria.

The ability to use oxygen gave the cells a boost in energy that they had. Some cells developed a flagella-- a whip-like addition that helped to move around and could metabolize or reproduce with unprecedented speeds. This vastly improved the powers of movement, metabolism, and reproduction. The result of the cells' new ability to utilize oxygen, which used to be a deadly poison, to convert nutrients into huge quantities of energy.

As cell division sped up, so did the process of mutations. Aerobic organisms (organisms that can use oxygen) began to populate the world more and more.

Eukaryotic Cells

As organisms became more complex, they evolved new ways of extracting more energy from their environment. Some even developed the ability to extract more complex body structures. Now, "the stage was set for the birth of the eukaryotic cell, [the cell with a nucleus]" (Transitional College of LEX). Eukaryotic cells evolved the structure of hundreds of membranes divided in the interior of the cell into different compartments, each taking a different cell function.

Originally, Eukaryotic cells used to be anaerobic-- cells that cannot use oxygen. These lived in the oxygen-free environment and used primitive or inefficient metabolic pathways, like glycolysis. This took a bold step, in which they took in more prokaryotic cells-- cells without a nucleus could use oxygen. Prokaryotes were the ancestors of mitochondria.

Symbiotic Relationship

When eukaryotes took in oxygen using prokaryotes, it formed what is known as a symbiotic relationship, a mutual benefit to eukaryote and prokaryote. Prokaryotes and the descendants-- mitochondria, consumed pyruvate that was thrown by the eukaryotic cells. In return, the mitochondria produced a lot of ATP needed by the cell for energy and gave back to the cell. Through symbiosis, eukaryotic cells could leave its energy production to mitochondria inside of it. Now, the outer membrane of a cell can be put to different uses. One of these uses included effective communication with the outside environment.

Some eukaryotic cells were not content with just having mitochondria. They also took in photosynthetic bacteria which was enabling cells to extract energy needed from the sunlight. These were the ancestors of chloroplasts (organelles found in all plant cells). These cells used the sun's energy to synthesize ATP and sugar. Even though there is no found available in the surrounding environment, these ancestors of the chloroplast are able to produce organic compounds required by host cells from nothing but the sunlight, CO2, and H2O. With these photosynthetic cells, the first plant cells were born, signifying the birth of the first plants.

Summary: Basically, as organisms adapted to the environment, which also transformed that environment. Then it is forced to adapt to the new environment, eventually, discovering a new way to evolve through symbiosis. This caused more evolution, causing cells to evolve into multicellular organisms.

The difference between Multicellular and Unicellular Organisms

The human's body contains 60 trillion cells in general. Many of them are different types, including fat cells, bone cells, and fat cells. However, they all evolved from the same parent cells. Not only that, but they all contain the same DNA inside each cell

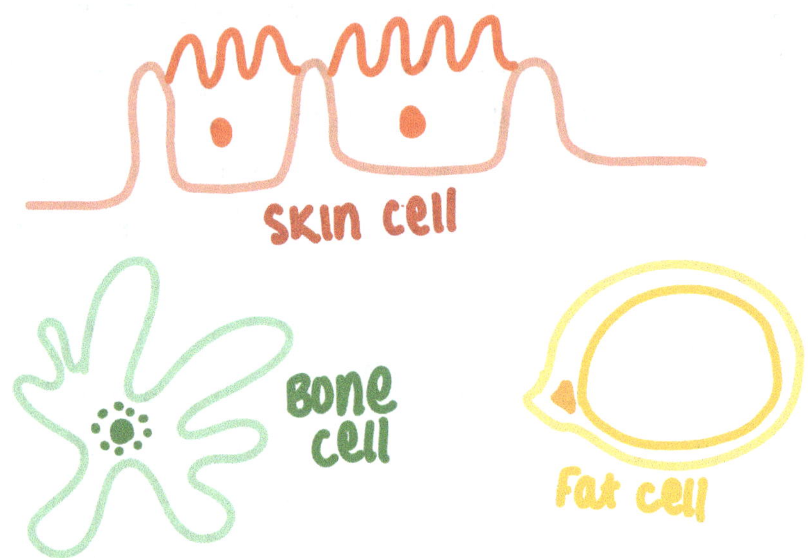

An experiment was conducted, in which a south African Clawed frog was experimented on:

The South African Clawed Frog Experiment

This experiment proves that one skin cell contains all of the DNA that is required to produce a complete individual organism. Meaning, all cells in an organism contain the same amount and type of DNA.

How do multicellular cells develop into different cells-- skin cells, eye cells, etc?

E. Coli v Multicellular Organisms

The difference between single cell organisms (EX used: E.Coli) differs from a cell in multicellular organisms through four main points.

1: Size and organelles

Multicellular cells are much bigger than E.Coli. Not only that, but multicellular organisms contain much more organelles. Organelles are compartments enclosed in their own membranes. They also perform special functions from the cell.

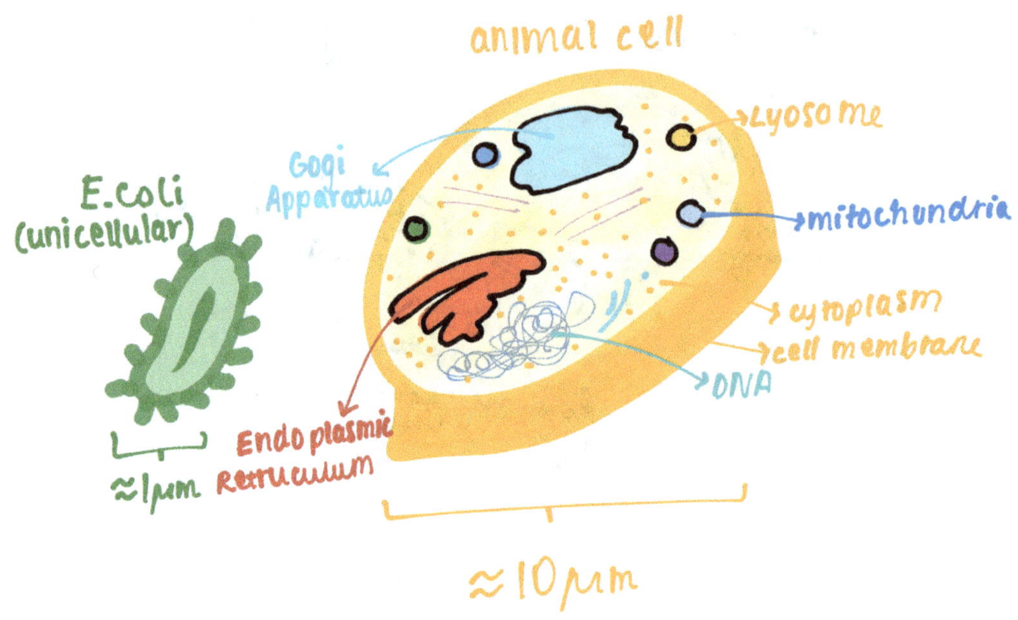

2: Cell membrane transport

E. Coli cells only have one layer of membrane separating them from the outside environment-- this transports materials in and out of the cell. Whilst, multicellular organisms not only have membranes separating them from the outside world, but they also have many types of membranes inside their bodies (EX; cell membrane, nuclear membrane). This transports materials in and out of the cells and within the cells inside the humans.

3: Introns and Spacers

E. Coli and multicellular organisms have different DNA. E. Coli are made up of genes-- sections of DNA that form codes for protein synthesis. In multicellular organisms, only 25% of DNA in humans and multicellular organisms consists of genes. 75% contains spacer DNA-- sessions of DNA that do not code proteins. Even today, scientists do not really know what spacer DNA is really for. Also, the 75% of DNA in multicellular organisms is a spacer and of the 25% of remaining genes, most of them are introns. In conclusion, this means that 98.5% of DNA are spacers or introns-- and scientists do not know what they are for. Actually, only 1.5% of the entire DNA is only put in use.

4: Cell division

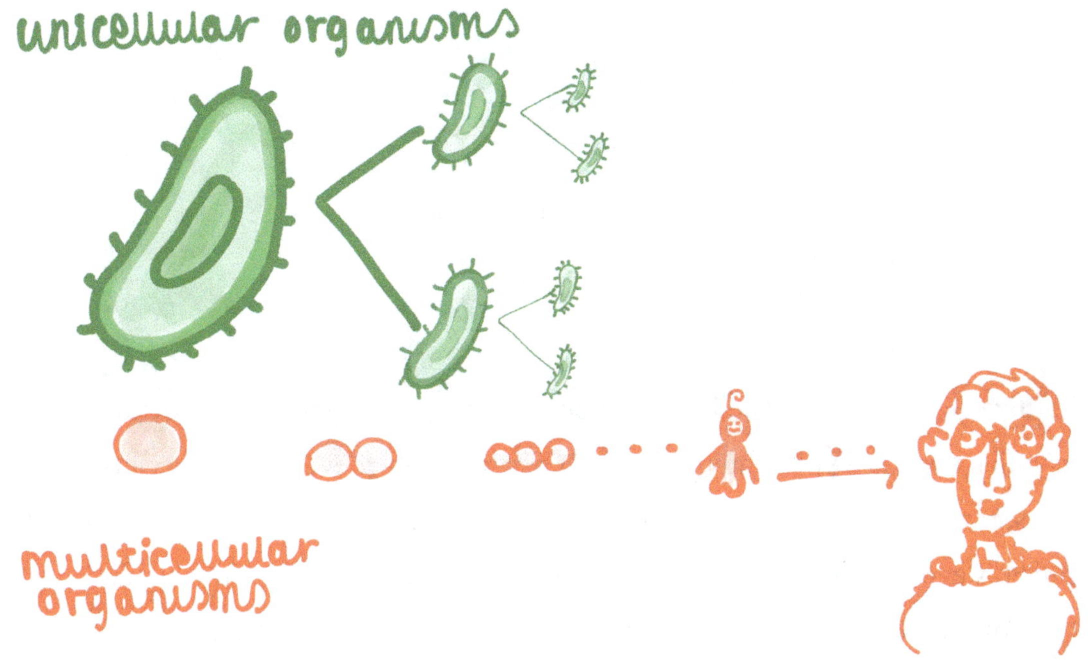

When E. Coli cells divide, each new cell is a complete, self-sufficient cell that can work perfectly on its own. Multicellular cells have a more complex method. When the fertilized egg cell of multicellular organisms divides, the new cells DO NOT disperse. Instead, they stick together to produce an individual organism that grows later into an adult. This process of multicellular organisms is known as development.

Development Process

Originally, multicellular organisms would start out as an egg. Up to a certain point, even after cell division, the organism would still look like little clumps of cells. Gradually, the shape would change. For example, the cells would form into a tadpole and develop then later into a full frog. Originally, the development of organisms in the beginning of development/evolution looked very similar.

Question: Development begins with Fertilization, but what happens before?

A 'macro' view of development starters with an overall big picture. Before fertilization, the determination of heads or tails occurs.

We will be using frogs as our example for this.

This is an unfertilized egg of a frog. The inner content of the egg is divided into two parts: the bottom half, in which there is a yolk-- an important substance in the development because it provides nutrients, is the vegetal pole-- the yolk end of the egg. This part of the egg becomes the tail of the organism's eventual body. The top half is known as the animal poe, in which it later becomes the head of the future body. This head to tail axis of the egg is known as the anteroposterior axis.

The egg predetermines the eventual head to tail axis of the body even before it becomes fertilized. What later is determined is the back and the belly.

Fertilization begins with the sperm cell entering the egg. Entry of the sperm causes the creation of the layer of the egg cell to rotate part way around the core of the egg. This is known as the cortex. This causes the animal pole to shift toward the point of sperm entry. The inner content of the egg also shifts at the same time. This movement is known as morphogenetic ("shape-forming") movement.

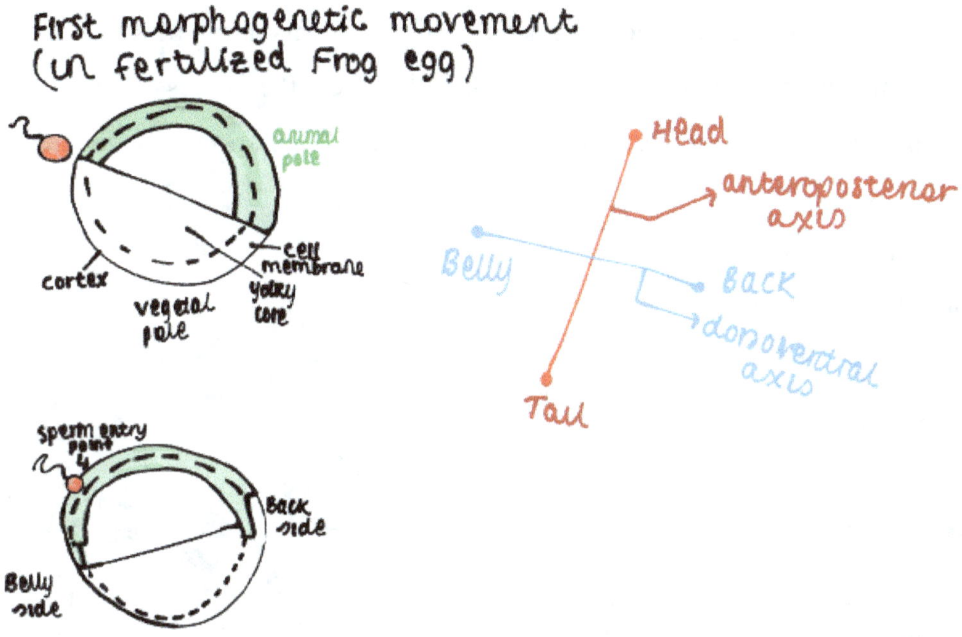

The sperm entry point becomes the location of the belly, and the opposite side of the egg becomes the back of the body. This back to belly abis of the egg is known as the dorsoventral axis.

Gastrulation

So, the anteroposterior axis is determined by the egg before it is fertilized, and only then, does the sperm fertilize the egg, determining the dorsoventral axis of the body. Once the egg is fertilized, it starts to undergo what is known as cleavage (or cell division).

Our examples that we will be looking at are the frog and the sea urchin.

First starting with Sea Urchin:

Sea Urchin

The cleavage begins with one cell and continues so until it forms into balls of cells, known as blastula. Sea urchins blatulas contain around 1000 cells. Then, the process of a series of changes occurs inside the blastula, in a process called gastrulation. Gastrulation is performed through three steps:

1: Outer cells start to tuck in

2: Cells inside the cavity grow

3: Forms the mouth and the anus

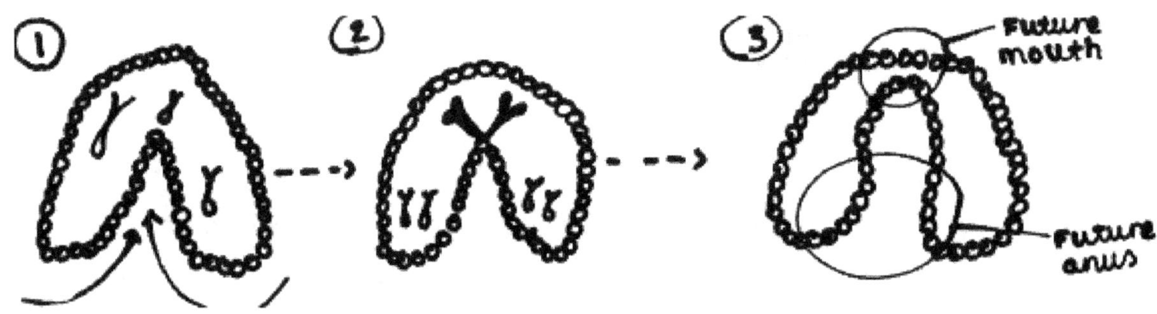

Gastrulation creates the gut tube, which is eventually forming into the digestive tract. This part is located from the mouth to the anus, down the middle of the body. The first thing that forms in the Sea Urchin is the digestive tract. During the course of gastrulation, cells of the embryo divide into three layers: inner endoderm, outer ectoderm and mesoderm (in the middle). Ectoderm is made out of the cells that started out on the outside of the blastula. These cells will eventually form into the skin and the nerve cells of the body. The endoderm consists of the cells that moved inwards during gastrulation. These, as said before, will form into the digestive tract, and lastly, the mesoderm is the layer of cells between the ectoderm and the endoderm. This will form the muscles and the connective tissues. After all of these formation occurs, the cells of the embryo will start to differentiate into special roles.

Differentiation of the Frog (Xenopus)

The differentiation moves from the general to specific.

In development, the cells divide over and over, however every cell contains exactly the same DNA, guaranteeing if the frog is indeed a frog. Then, once the cells develop a general outline, the cells start to differentiate into specific roles for the body. This process always starts with the most important functions of the frog.

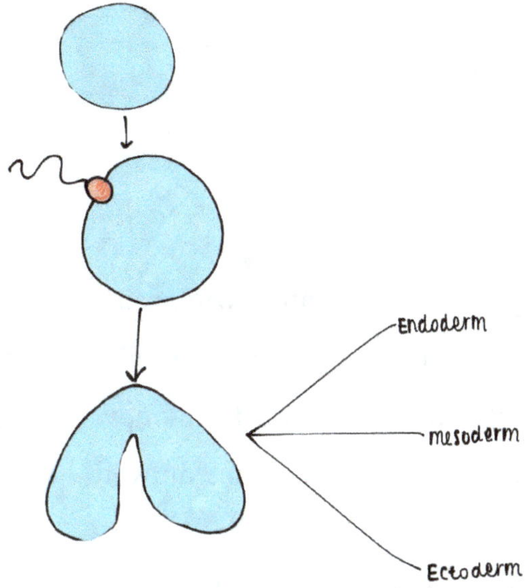

In these cases, the endoderm would determine the anus, bladder, intestine, liver, pancreas, esophagus, stomach, lungs, windpipe, ears, and thyroid gland. The ectoderm would determine skin and nail, hair, eyes, nerves, and the brain. Mesoderm, will form the bones, kidneys, muscles, blood cells, blood vessels, and the heart.

How do cells differentiate?

Development grows from the big picture into the details. This is when the cell starts to develop individual cells' differences. The way cells decide what they will perform, it occurs with the help of the fertilized egg:"As a fertilized egg divides from one into two and then 4,8,16 cells, cells aren't just dividing, they're also different".

When do cells actually begin to acquire different roles?

When a fertilized egg of a mouse got to the point of diving into 8 cells, the cells were separated and were inserted into the uteri of 8 different female mice. The result would be the birth of eight normal mice. But, if the mouse cells are separated after the egg is divided into sixteen, the result would be that no mice shall be born. Up to the eight cell stage, every one of the cells in the mouse egg are totipotent, meaning they have the ability to develop into any part of a complete organism. But once divided into the sixteen cells, the individual cells lose their ability to differentiate. In other words, differentiation of the cell roles begins around the time when the egg divides into sixteen bells. The difference between division into eight cells versus the division into sixteen cells, have to do with the difference in interaction that takes place amongst the cells themselves. Cells of multicellular organisms are not just connected.

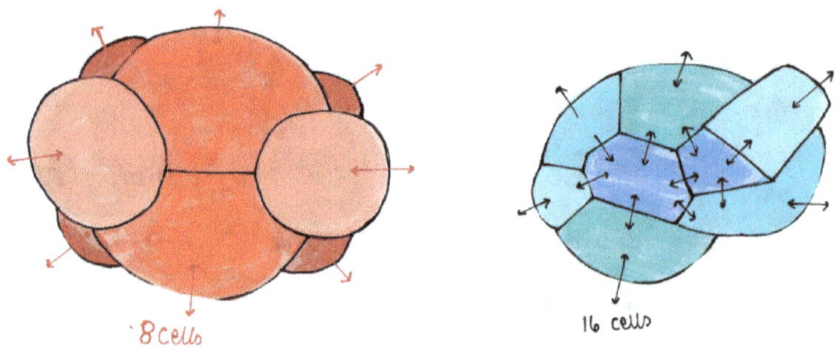

The cells are constantly interacting-- exchanging the information with each other at several levels. The nature of this interaction changes when the 8-cell embryo divides into the 16-cell embryo. When there are only 8 cells, every cell faces both inward and outward. When there are 16 cells, some cells face only inward and other cells only that face outward.

Information exchanged between cells then begins to vary, depending on whether it is outward or inward facing. This has a big impact on what the future function of each cell will be. However, nobody knows what the exact timing is within different organisms. Using our example on mice, mouse clones cannot be produced from embryos that are beyond eight cell stage, but cows and sheep can be cloned from 16-cell embryos or even 32-cell embryos. Scientists have already put the knowledge to practical use, by cloning cow embryos to produce better milk, beef, etc.. As for the determination part, whatever will determine a cell will become a part of the human body early.

Cell determination, memory and differentiation
People can tell when determination takes place. By transplanting cells from one region of developing embryos into the different region. For example, through using amphibian embryos, the piece of tissue from the region that will become skin might be transplanted to the region that will become the brain. If transplanted cells have not been yet determined, they will develop into a brain cell, just like the cells around them in the "new home". From this the development of the embryo continues in the normal fashion. But if the translated cells were already determined as the skin cells before they were transplanted, they will continue to develop and differentiate into the skin cells, even though the cells around them differentiate into the brain cells. Scientists tried this experiment at different stages of development of animal embryos to figure out just when

determination occurs in the body. Experiments have shown that when a cell is 'determined', it memorizes what it is supposed to become. This phenomenon is called cell memory. Scientists used the word determination to describe when the cell becomes committed to developing a certain way (even before the sign of that development part). Whilst, differentiation is when a cell actually starts changing in the visible way. This is the difference between determination and differentiation.

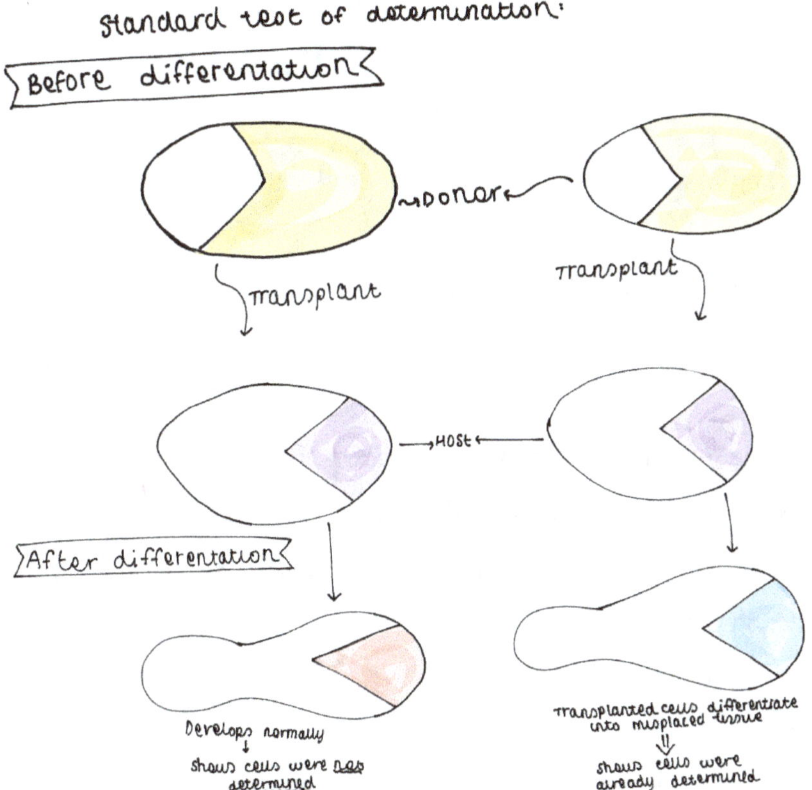

The Chicken Experiment

The Chicken Experiment was performed, having the goal to transplant a piece of tissue inside the egg during development of a chick embryo. By observing development after transplant, scientists would be able to tell how cells differentiate into the future roles. On the third day of incubation or the period in which the hen sits on the egg until it hatches, the data collectors check if the embryo hasn't developed any real wings or legs as of yet, but the places where there will be wings and the legs have already been determined. So, if a piece of tissue from a part that is already determined to become a thigh is transplanted to the tip of a part that is determined to become a wing, what would happen?

The wing would grow a toe! The experiment showed that the transplanted tissue had already been determined to become a part of the leg, however, it had not yet been determined to become a definite thigh. It only had to 'become a leg' (with no specifications). That 'become a leg' tissue was transplanted to the region where the cell will be surrounded by 'become a wing' cells. The cell would then pick up this information and conclude that it must become a tip of where it is. Usually, the cells would turn into a 'wingtip', but instead, it becomes developed into the 'leg tip' (a toe).

Cell to Cell Interactions

Let's take a look at how interactions among cells in a multicellular organism cause each other to differentiate.

There are two types of cell interactions: local interactions and remote.

<u>Induction (Local interactions)</u>

Local interactions between adjacent cells are known as inductive interactions.

Going back to the example with the frog Xenopus, if the blastula develops normally, then gastrulation should produce the endoderm, mesoderm, and the ectoderm. Let's see what happens when blastula's cells separate from each other.

<u>Mesoderm Induction in the Xenopus</u>

If separate embryo cells are placed in a culture, then each of the cells will develop individually. Some will develop features of the ectoderm, others may of the endoderm, but none will develop features of the mesoderm.

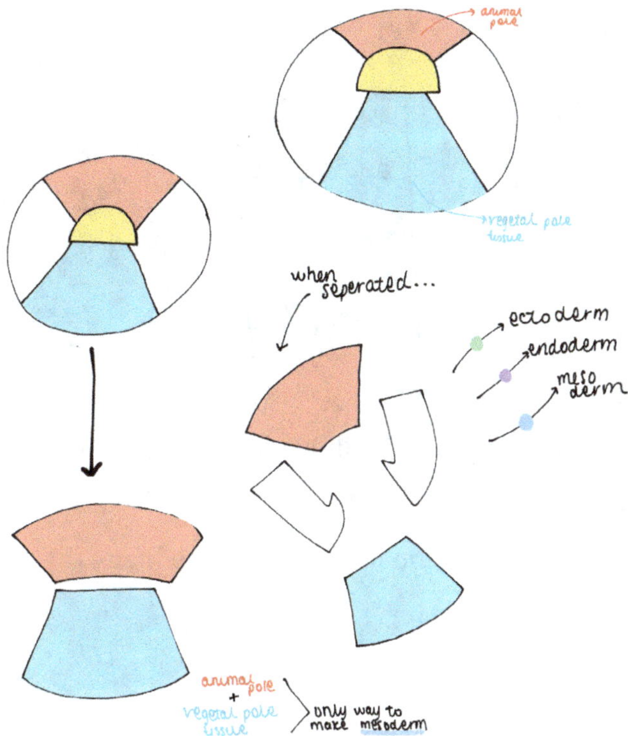

If you put an early Xenopys embryo in a solution that contains no Ca2+ or Mg+2 ions, the cells lose cohesiveness and may be separated!

So why don't any of them turn into mesoderm when you separate them? When after extracting all except for the combined animal poke and the vegetal pole combined, and after having that cultured, it will mainly produce mesodermal tissue. So, if you stick the tissue from the animal pole next to the tissue from the vegetal pole-- most of the animal pole tissue would turn into the mesoderm. Induction is when development of the group of cells is influenced by the cells next to them.

These are simpler diagrams of finding and creating the mesoderm:

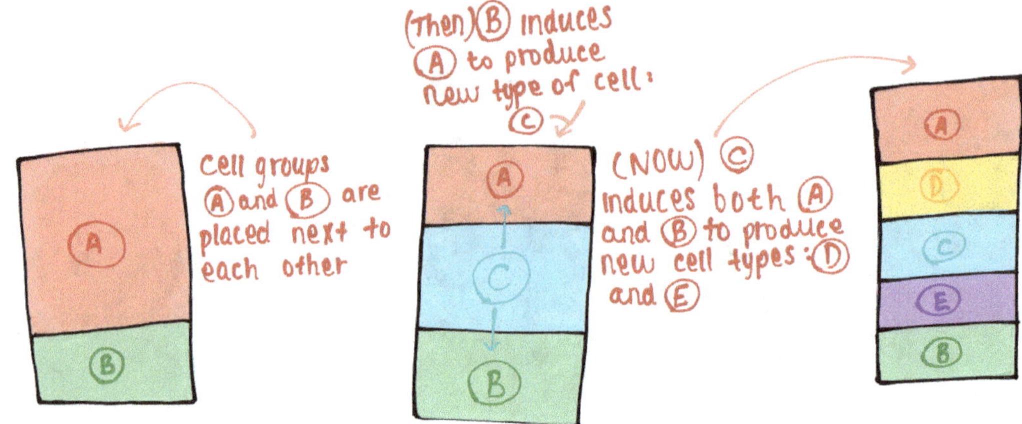

Basically, the process that starts with induction between two types of cells, in this case A and B, ends up with five different types: A, B, C, D, E. This is known as sequential induction, which is a process of a repeated induction that can produce many different types of cells from only just a few.

Morphogen Gradient (Remote interactions)

Another type of interaction is known as morphogen gradient. Unlike induction, where the interaction happens between adjacent cells, the morphogen gradient is an interaction that affects an entire embryo when developing what happens: a patch of tissue somewhere in the embryo that produces a substance that spreads slowly from one point through a neighboring tissue. As it spreads, the substance (known as morphogen), reaches other cells that are at different levels of concentration (higher levels meaning that it is close to the source, while low levels are spread further away). This change in concentration that is based on distance is known as a gradient, which is based off of the term, morphogen gradient. The morphogen gradient acts as a signal to each cell. Scientists have not yet identified these substances. Many morphogens are actually proteins– for example, activins and morphogenetic proteins. But, there are a few non-protein morphogens as well, like retinoic acid.

If the cell gets a high concentration of morphogen, it would know that it is close to its source. However, if cells get a low concentration, it knows that it is far away from the production spot of morphogen.

Here is an example of a fruit fly's embryo:

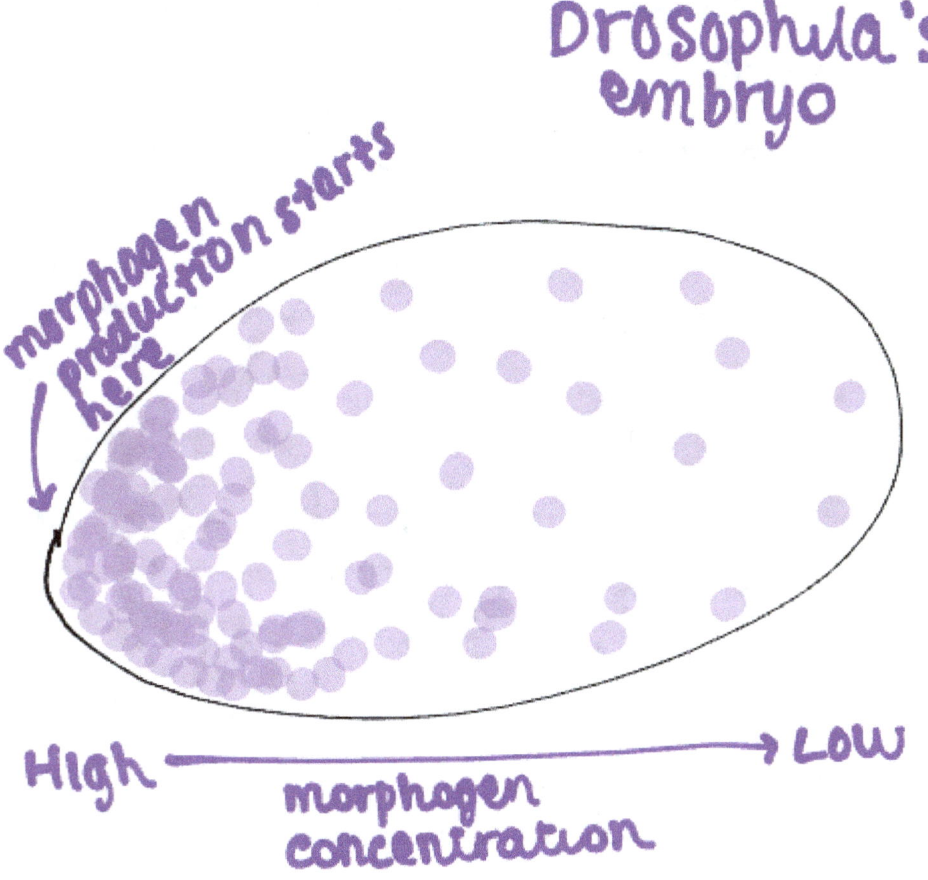

Many other interactions exist besides induction and morphogen gradient. Through these interactions, the cells of developing embryos soon acquire different features as well as functions.

So how does development and DNA connect with each other?

To answer this question, let us use the example of the Drosophila (fruit fly).

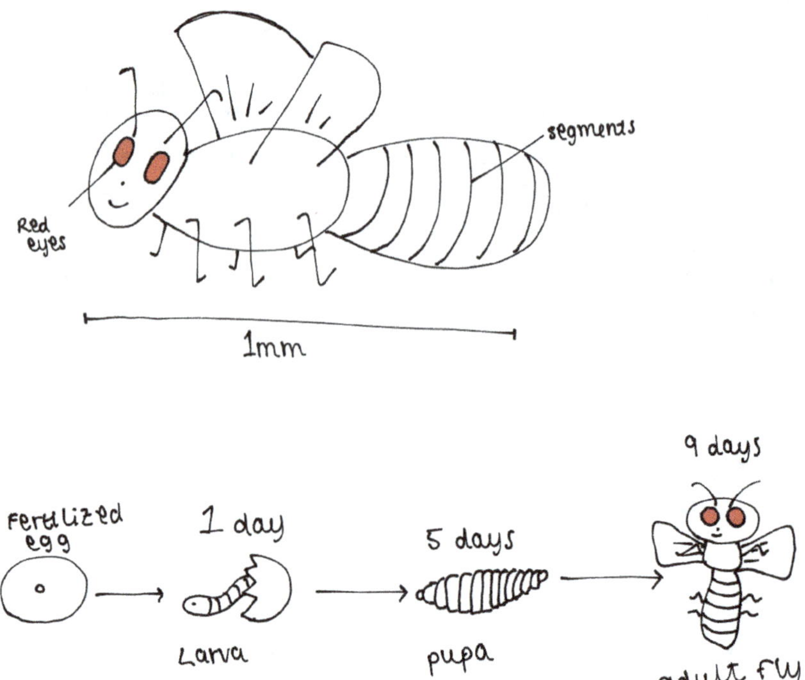

Fruit flies in general are a very commonly used creature to perform experiments about development, due to the ease in which scientists can find information about the matter. In multicellular organisms, the development process moves on as cells in the embryo perform complicated interactions with each other. It is difficult to track this relationship of every change in every cell to its own DNA (or genes). To clear up the confusion, scientists studied the Drosophila in detail to trace these changes. The reason for choosing this species is because it takes them only nine days to perform their development period from fertilization to adulthood. The drosophila egg hatches a day after fertilization, then becomes a pupa by day 5 and then is a full adult by day 9.

Most of the fly's body is made of segments. By studying how these segments are formed, biologists learned how development and DNA are connected together. So, the reason in which biologists are interested in the genes that determine the segment pattern of a fruit fly is because of the fact that drosophila genes are very similar to the genes that control the development of the human body. They can give out clues on how the development process works in humans, frogs, and any other multicellular organism.

After the one day fertilization period of the egg, the drosophila goes through the process of embryonic development.

The drosophila does not undergo cell division right away. At first, the nucleus divides repeatedly to produce the mass of cytoplasm with many nuclei (known as syncytium) (A). Then, multiple nuclei migrate toward the surface of the egg, which forms layers known as syncytial blastoderm (B). Lastly, the membrane will grow inward from the surface of the egg to enclose each nucleus. This forms a cellular blastoderm that consists of 6000 separate cells (C).

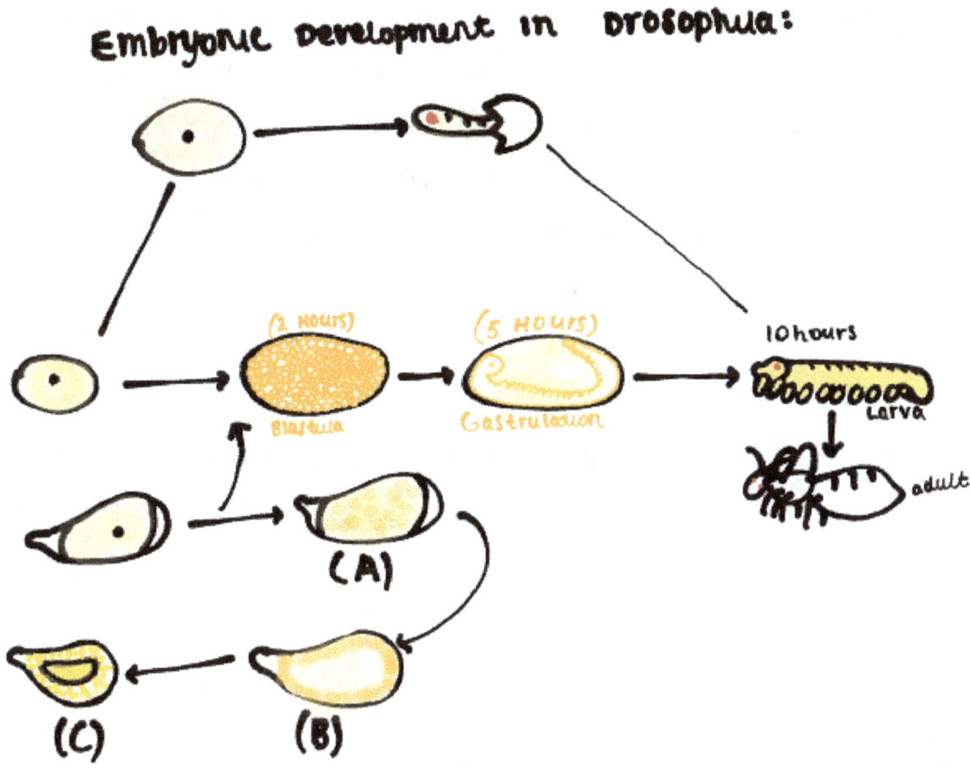

After cellularization is complete, gastrulation will begin.

Five Gene Groups

During the embryonic development of the Drosophila, there are five gene groups that determine the formation of segments. A gene group is a set of genes to which humans assign a single name, due to the fact that they produce similar effects at a certain point in development. The five gene groups are:

- Egg-polarity genes
- Gap genes

- Pair-rule genes

- Segment-polarity genes

- Homeotic Selector genes

These five genes have a very interesting relationship with each other.

The description of how scientists figured out which genes are related to which segments of the Drosophila embryo. Scientists looked for fruit flies whose genes had been activated-- by a mutation. Then, they checked to see which segments were affected by that certain mutation. That told them which segments are dependent on genes for formation. So basically, they would learn about the gene's importance by seeing what happens when it is mutated.

Let's look at the second and third gene groups that we mentioned in the list: Gap genes and Pair-rule genes when they are mutated.

Mutations of Drosophila: Gap Genes

If a gap gene that is known as Kruppel mutates, then the larva would develop with the entire midsection missing! This shows that the kruppel gene is crucial to the development of middle segments of the larva.

Mutations of Drosophila: Pair-rule Genes

If a pair -rule gene known as the Fushi-Tarazu disappears, then mutates, these sections would go missing. As a result, the larva would develop with all od the odd numbered segments missing, showing for the scientists that Fushi-Tarazu genes are crucial for the development of odd-numbered segments

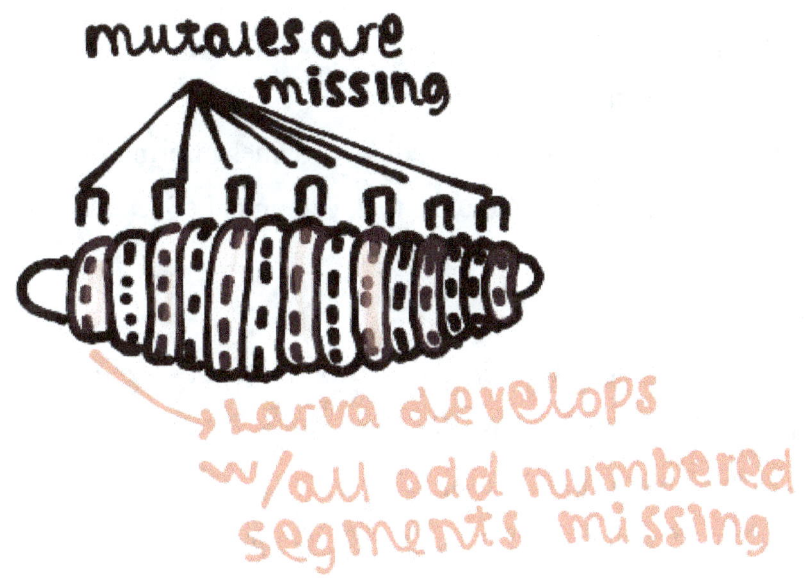

For both of these genes-- the gap genes and the pair-rule genes– there is an order of priority in the expression of genes. Here is the order of the gene group that shows what each gene determines and which one is a higher priority to keep. *Notice how the order for prioritization goes from general to specific:*

1. **Egg polarity genes:** Determines the front and the rear of the egg
2. **Gap genes:** Determines the middle versus the rest of the embryo
3. **Pair-rule genes:** Determines the odd and even segments
4. **Segment genes:** Determines the front and reach for each segment
5. **Homeotic Selector genes:** Determines the final position of the cell

So, what is gene expression?

When a gene is expressed, it means that the gene is doing its job. During development, the cells of the embryo can tell where they are located from the concentration of substances, known as morphogens, at different spots in the embryo. In embryonic development of the Drosophila, morphogens are proteins.

When a gene goes to work, it is producing a certain type of protein. That protein spreads throughout the entire embryo, defining the positions of its cells. When there are high concentrations of protein, they will be closer to the origin. And when there are low concentrations of proteins, it is far from the origin.

This way, the protein's concentration gradient provides information to all cells in the embryo, telling them the final position and ensuite development of segments in their proper place.

How does the development of segments relate to the distribution of proteins in the embryo?
We will use the Drosophila example to answer this question.

The Embryonic Development of the Drosophila

We will break down each type of the five gene groups and look at their embryonic development:

- **Egg-Polarity Gene Group** (including the genes Bicoid and Oskar):

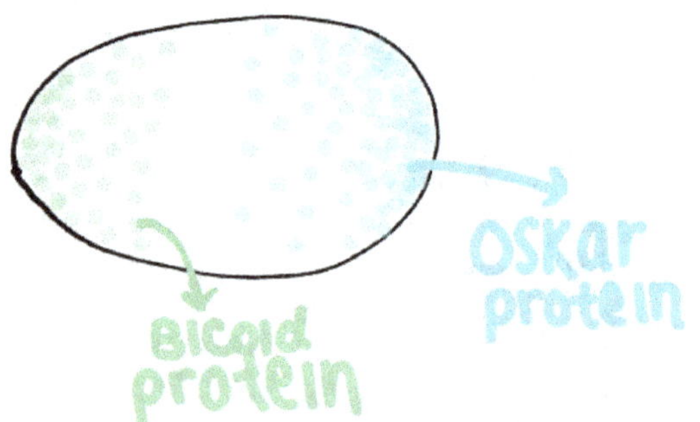

The Bicoid protein spreads from the front and the Oskar protein spreads from the back. Both proteins are scarce in the middle. So, the middle of the embryo is a part of a different protein production.

- **Gap-Gene Group** (including Kruppel and Hunchback gene):

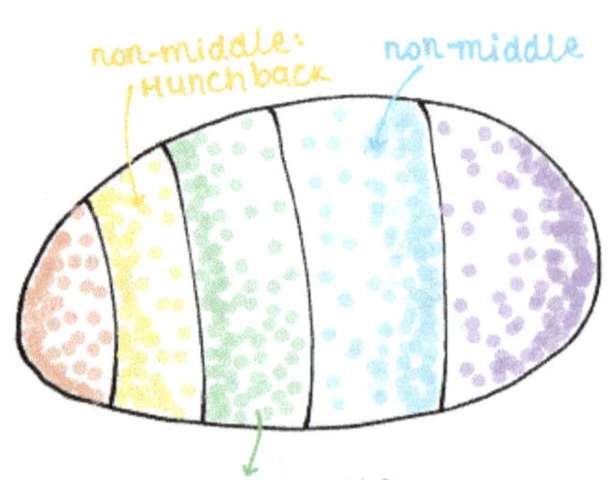

The Bicoid and the Oskar proteins stimulate the expression of gap genes in the middle. And, the non-middle parts of the embryo are producing the genes Kruppel and Hunchback.

Pair-Rule Gene Group (including the genes even-skipped (Eve) and the Fushi Tarazu (Ftz)): Different concentration gradients or proteins are produced by the egg-polarity. Gap genes stimulate the expression of the Pair-rule genes in alternate segments. Now, at this point in time, the embryo has 6000 cells.

- **Segment-polarity Gene Group** (from (3)): Different combinations of proteins that are produced by the egg-polarity and all previous gene groups stimulate this. To better understand this, an example would be the pressure of the segment-polarity genes which are divided into four regions (as shown in the diagram). At this stage, the position of the individual cells are determined.

- **Homeotic selector gene group** (determines final positions of the cells!!!):

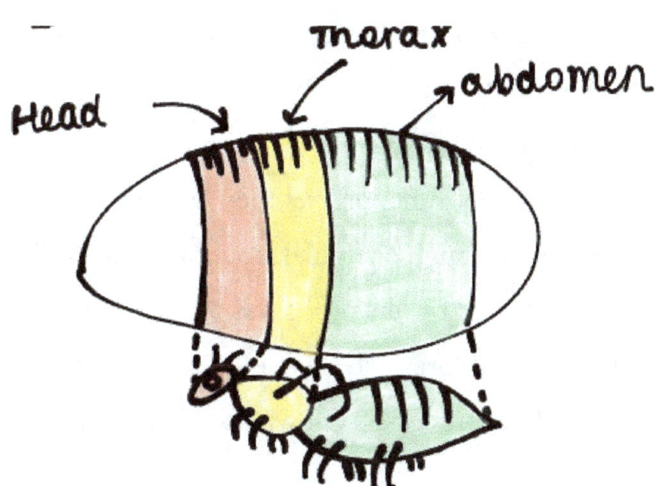

Homeotic selector genes implant in each of the cells a memory in which either the head, thorax, or abdomen is assigned to it. Each segment is determined by precious gene groups belongings.

Now, let us talk about proteins concentration gradients:

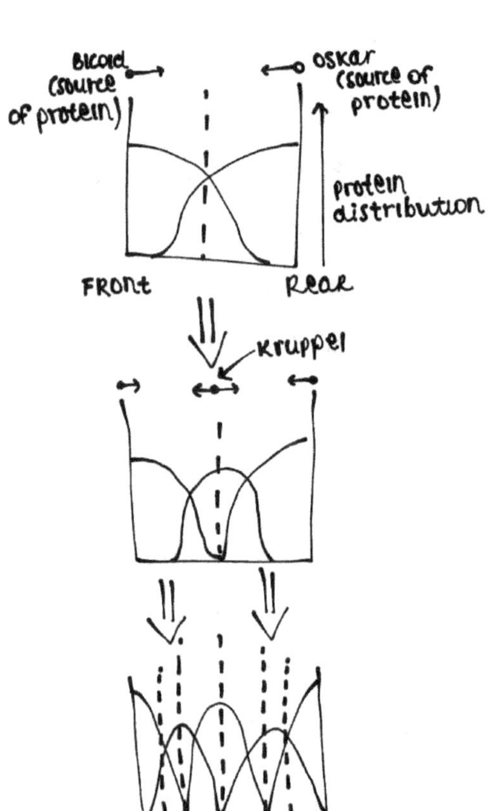

A protein is produced

Here, another proteins is produced in the middle

Now, the proteins are being produced in between

So we learn all of this, but how is gene expression controlled? What stops cells from saying "Me be no finger. Me be toe"?

How does the expression of individual genes get controlled in the course of an organism's development?

What controls gene expression?

Every cell acquired a different form and a different function, yet they have the same DNA in each cell. The body of an organism is formed through chemical reactions that synthesize protein from the DNA genes in every cell. The crucial part of this process is to control which genes in the DNA molecule (in a particular cell) will be expressed. Meaning, exactly what type of protein will be synthesized and how much of it? This process is known as the control of gene expression.

The Control of Gene Expression in Eukaryotic cells

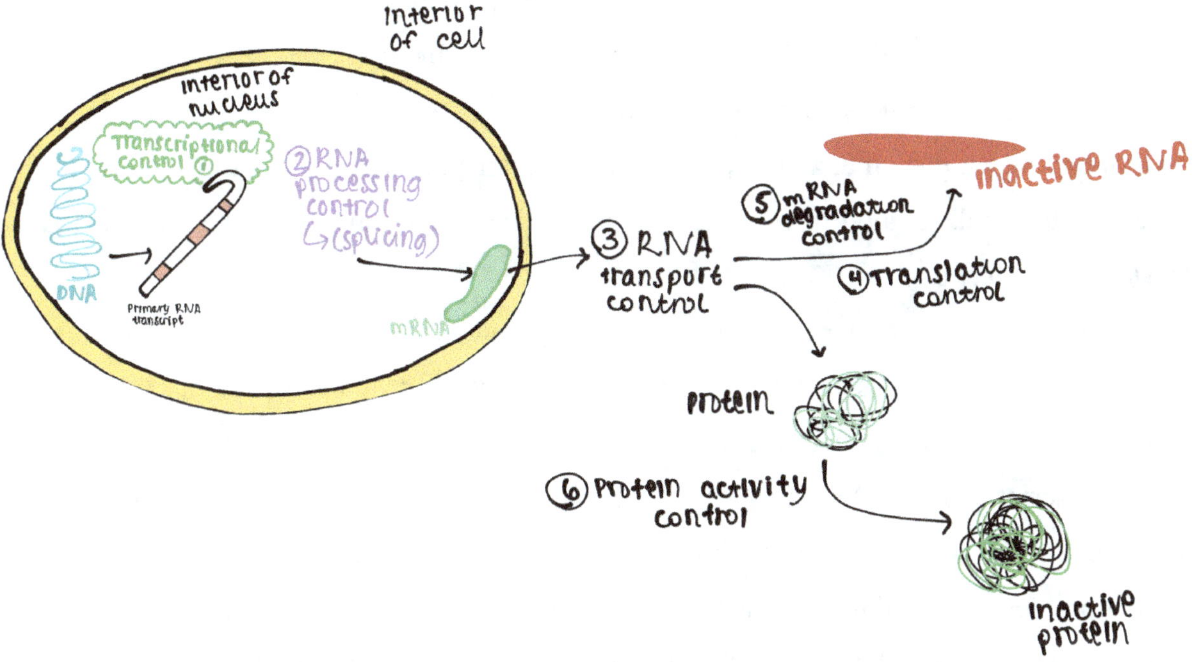

DNA produces RNA, and the control takes place between the steps shown in the diagram, which produces a specific protein. This is a six step process, the most important one being Step 1-- Transcriptional control. It is strange because transcriptional control takes place while the proteins are being synthesized, yet the proteins are being controlled at the same time. Let us take a closer look at these:

How proteins transmit information:

Proteins essentially provide the 'skeleton' that supports the structure of a cell. They convert nutrients that enter cells into different stage substances. Proteins serve as information. How? If they are not living creatures (Meaning: have no eyes, ears, legs, etc) and cannot move freely but must be assigned, how? By their shape! The way those gene regulatory proteins bind with DNA is quite interesting and perhaps extraordinary.

Proteins that are known as enzymes are involved in processes of protein synthesis, RNA and DNA replication. Some of these enzyme proteins include:

- RNA Polymerase-- copies genetic information from the DNA to the RNA as a part of proteins synthesis
- Spliceosome Complex-- is actively involved in the RNA splicing
- DNA helicase-- unravels the DNA helix during DNA replication
- DNA Polymerase-- builds new DNA strands

All of these examples of enzymes perform a particular function(s) when they bind with other molecules that make a good fit with their shape. Basically, the function of the proteins IS its shape (*form is a function*).

Certain proteins control gene expression by using their shape to bind with DNA in a certain way. These are known as gene regulatory proteins. Several proteins that do this, all of which play a crucial role in the process of protein synthesis.

The transcription of RNA from DNA begins with the binding of the DNA polymerase molecule to the section of DNA. This is known as the promoter-- labelled as (1) in the diagram below. The promoter is located next to the section of DNA base sequence for certain gene-regulatory sequences, determining whether the transcription of genes will begin or not. When specific proteins bind to regulatory sequences, it enables the RNA polymerase to initiate the transcription process.

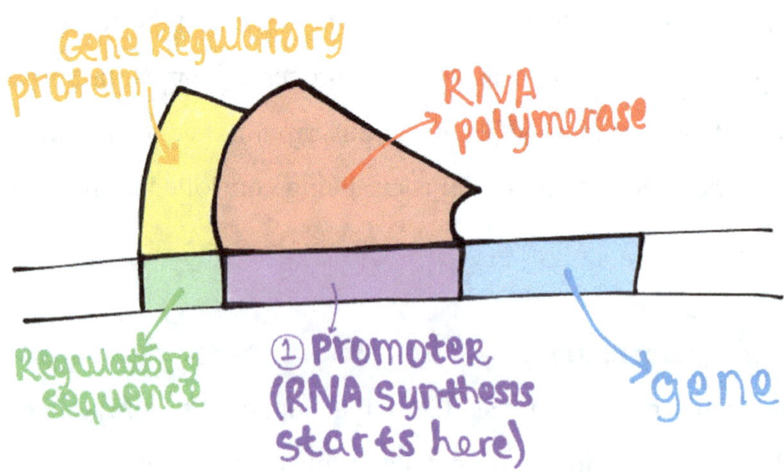

This is the start of the transcription with the binding of the gene regulatory protein to the regulatory sequence.

Each gene regulatory protein has its own particular shape. But so does the DNA base sequence and where the protein is to bind. This means that for synthesis of a given protein (let's say protein A), there is a specific regulatory protein base sequence and regulatory protein that is capable of binding it. Each base typically contains less than about 20 base pairs. Another protein (example: protein B) will have its own particular base sequence and correspond to another regulatory protein.

Essentially, the fit of a regulatory protein towards a DNA base sequence (depending on the shape) is what determines sequences and proteins that bind to these enzymes. Everything depends on the perfect fit between respective shapes of base sequences and regulatory protein permits. This is not just the general shape of each molecule, it has to fit down to each detail.

Here is an enlargement of the DNA base sequence:

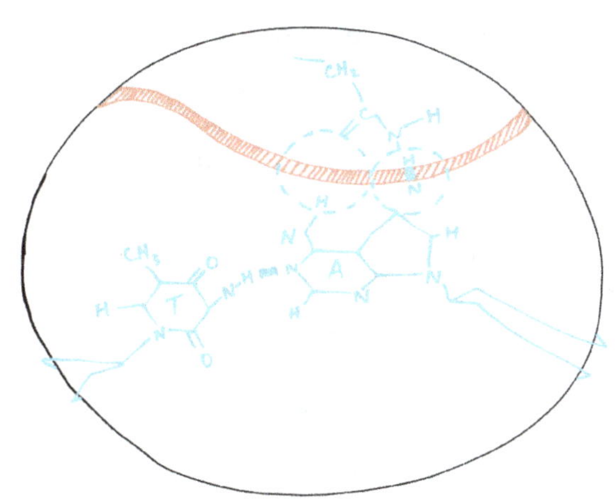

Here, the gene regulatory protein molecules are bound tightly to the DNA molecule at two sites, both of them being Hydrogen bonds. The DNA base sequence side chains have to bind to the side chains of amino acids that make up proteins.

The gene regulatory protein ban recognizes (from the outside of the double helix) specific DNA base sequences that can bind to. The DNA double helix contains what are known as major grooves and minor grooves. And, as said before, there are four bases of DNA.

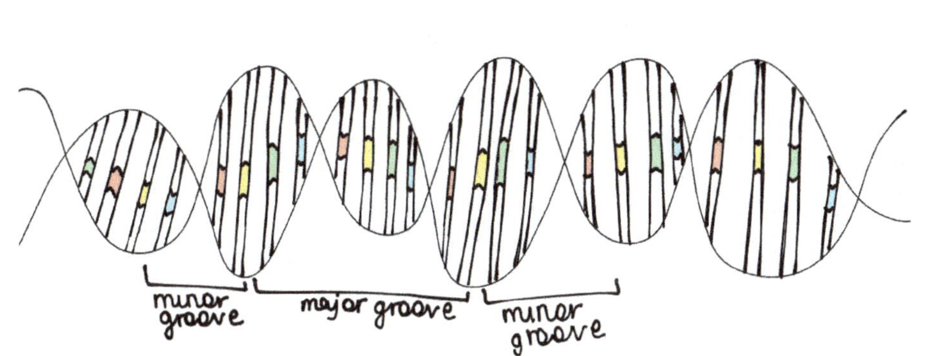

The edge of each base pair can be seen from the outside of the major and the minor grooves. However, only the major groove pattern of each base pair can be distinguished from the other three base pairs. Gene regulatory proteins usually bind at the major groove, because that is where they can identify unique patterns for each base pair. Due to the patterns on the edge of the major groove, gene regulatory proteins do not have to force open the double helix to see what to bind to.

On the big long DNA molecule, it seems that it should not be easy for a specific regulatory protein to encounter specific regulatory sequences. To make this process easier required the use of certain common structures found in most of the regulatory proteins the α helix and the β sheet.

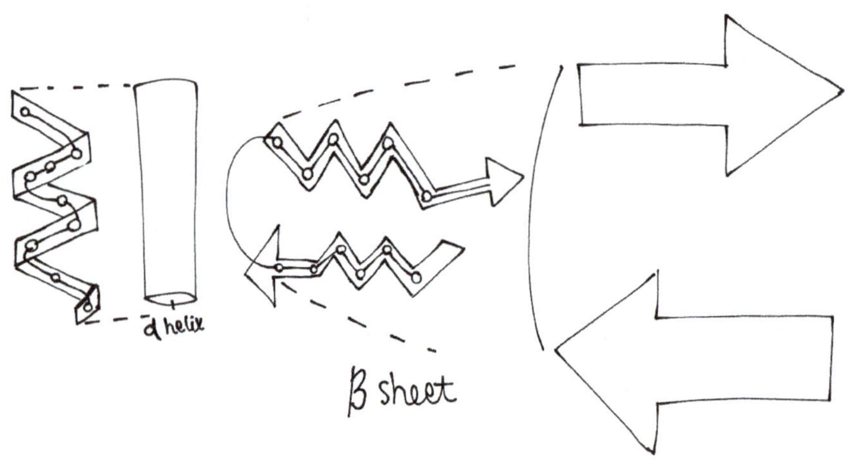

The left side of each illustration is a 'sketch' of each helix and sheet of how both α helix and β sheet consists of a series of peptide bonds that are linked together. The right side of each illustration shows how the structures are commonly drawn in diagram form.

This α helix and β sheet combine together to form a type of DNA binding structure known as motifs. Each motif binds to a specific regulatory base sequence of DNA. This forms a pattern in which DNA and protein are bound together.

This diagram is a helix-turn-helix motif, which is made out of three α helices. This motif fits perfectly into only a specific DNA base.

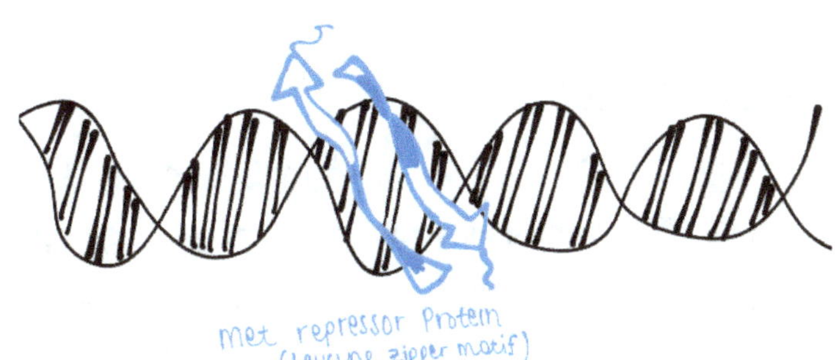

This diagram is of the Leucine zipper motif, made out of two β sheets.

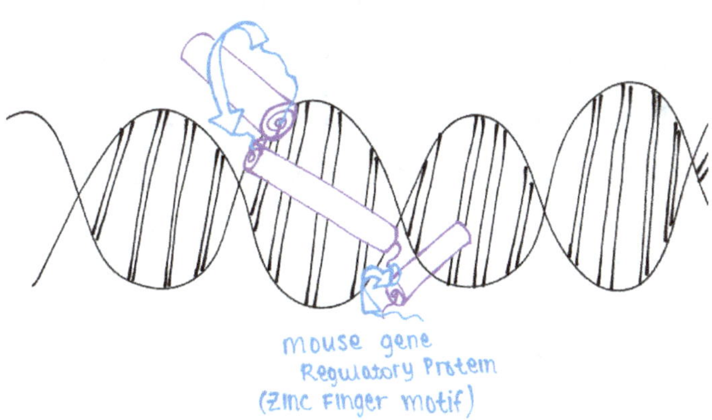

This diagram is of the Zinc Finger motif. It is known as this due to its combination of one α helix and one β sheet being held together by the atom called zinc. The protein that is shown uses a pattern of three zinc fingers to bind to a specific DNA base sequence.

For a quick summary, essentially, everything depends on the shape of proteins and the DNA. When the shape fits perfectly, it allows for genetic information to be transmitted, and is able to control whether or not the gene will be expressed or not.

To look closer at how gene expression is controlled, we will examine the most important type of control-- the transcriptional control.

A genetic switch is the ability to describe a combination of gene regulatory protein with a specific base sequence. The "switch" part of the gene switch refers to the image of the control mechanism machine.

When proteins bind to genes, they act as a switch that determines whether or not the gene will be transcribed or not. The switch is turned ON or OFF by a certain process. In fact, two types of gene regulatory proteins do this:

- **Activators:** which promote transcription
- **Repressors:** which prevent transcription

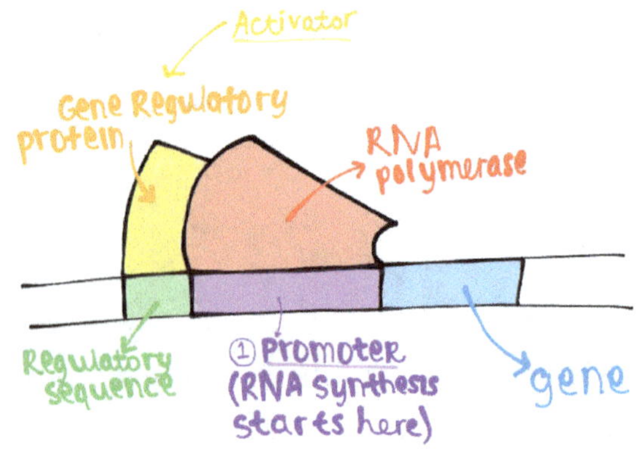

The Activator:

The Activator helps the RNA polymerase bind to the RNA synthesis starting site. Specifically, this protein assists the RNA polymerase that is unable to find the gene on its own, or if it does find the gene that is too weak to start the transcriptional process, then the activator protein enables the RNA polymerase to transcribe the gene. This permits protein synthesis. Since the bind of the activator causes for the DNA to switch the gene to ON, it is known as the positive control.

The Repressor:

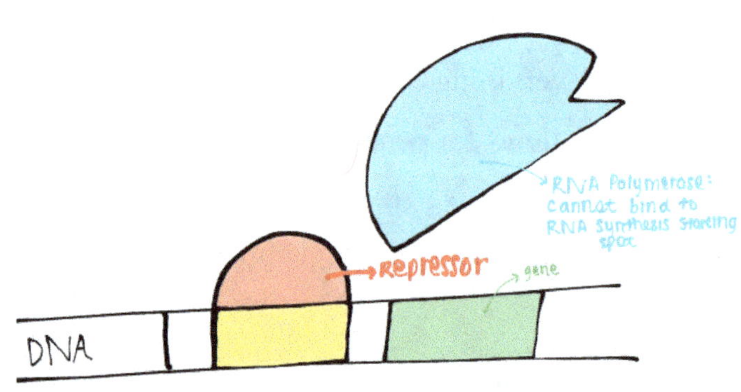

The Protein that Prevents Transcription

The repressor binds to the starting spot for the RNA synthesis, preventing the RNA polymerase to bind there, on top of the promoter. The repressor protein prevents the transcription of the genes and so, no protein will become synthesized. Since the binding of the repressor to DNA switches the gene to OFF, it is known as negative control.

Now, let's look at how gene expression is controlled in Unicellular organisms (using E.Coli as our example). Unicellular organisms work differently than multicellular organisms, so it is important to look at how they work and compare the two later.

To see how gene expression is controlled inside a unicellular living cell, we will look at what happens inside what is known as the E.Coli bacterium.

The reason for using E.Coli so much as our example for unicellular organisms, is due to it being the first organism in which biologists were able to see how the idea of control actually takes place. The scientists who were the first who did this experiment were Francois Jacob and Jacques Monod:

Jacques Monod was born in Paris, France in February 1910. His father was a painter, however most of his family were doctors. His mother was American, but with a Scottish descent. WHen studying Monod studied at the lycee de Cannes, and studied under Monsieur Dor de la Souchere, who was the founder and curator of the Antibes museum. Monod developed in his private life a Greek culture, though he was not Greek himself, becoming a very spiritual person during his younger years. Monod's interest in biology came from his father, who spoke of and introduced the laws of Darwin to him. In 1928, he registered for a degree in Natural Sciences (during that time, it ook about twenty years to accomplish). During his time studying, he believed that his discoveries and research was owed to other great biologists: Andre Lwoff, Boris Ephrussi, and Louis Rapkine. Monod got his Science Degree in 1931, and his doctorate in 1941. He spent some time at the California Institute of Technology and joined the Institut Pasteur. He was made the Director of the Cell Biochemistry Department in 195, and was later appointed as the professor of the Chemistry of Metabolism at the Sorbonne. In 1967, he then became a professor at the College de France. During his times studying, researching, and lecturing others, he met with Francois Jacob, another french biologist, and together originated the idea of the control of different enzyme levels in all cells during transcription. They used E. Coli to do this. Together, they shared the 1965 Nobel Prize in Medicine with Andre Lwoff.

Gene-Regulatory Proteins Inside E.Coli

E. Coli mostly eats glucose-- a type of sugar. If glucose is available near the E.Coli, it will take the sugar inside the cell and extract the energy that it needs. However, there is no guarantee that glucose will always be near E.Coli. If there is no glucose, then the E.Coli will die (unless there are other nutrients nearby). E.Coli may also eat another type of sugar-- lactose. However, it cannot eat lactose like it would eat glucose. It first has to break down and convert the lactose into glucose. E.Coli has a special gene for this to be done. But, it does not have to make that enzyme all the time.

If the environment surrounding the E.Coli has lactose but no glucose, then the "lactose-breakdown" enzyme will act and break it down (and thus be synthesized from the DNA of E.Coli).

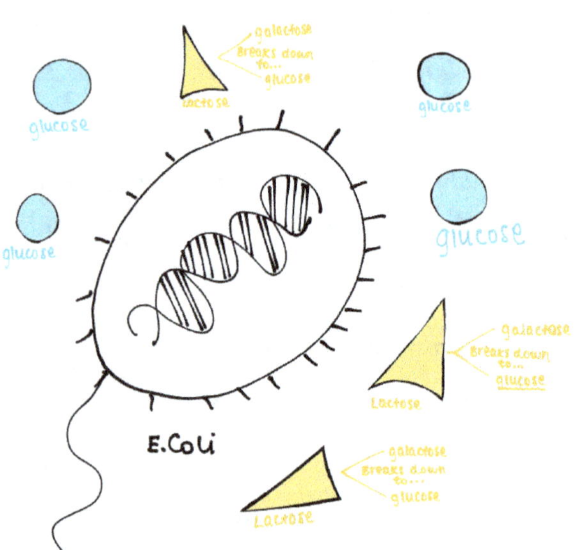

If there is any glucose in the environment, then no matter how little and no matter how much lactose there is available compared to glucose, it will hardly use the "lactose-breakdown" enzyme. Basically, the presence or absence of glucose or lactose controls the expression of a gene that codes the lactose-breakdown enzyme, turning the enzyme ON or OFF.

So how does the synthesis of the lactose-breakdown enzyme work?

The big idea of this is essentially that only when there is lactose and no glucose, then the RNA Polymerase begins the transcription process, resulting in the synthesis of the lactose-breakdown enzyme. This means that the E.Coli will synthesize the lactose-breakdown enzyme only when it is truly needed (when there is lactose but no glucose at all).

This synthesis can be seen based on the type of regulatory protein that is binding to DNA-- the activator or the repressor. This determines whether the genetic switch is ON or OFF as well. Not only that, but the type of protein binding to DNA is determined by the environment surrounding the E.Coli, whether there is glucose or lactose.

As said before about the multicellular organism's ability of gene control, that is controlled by binding proteins to specific

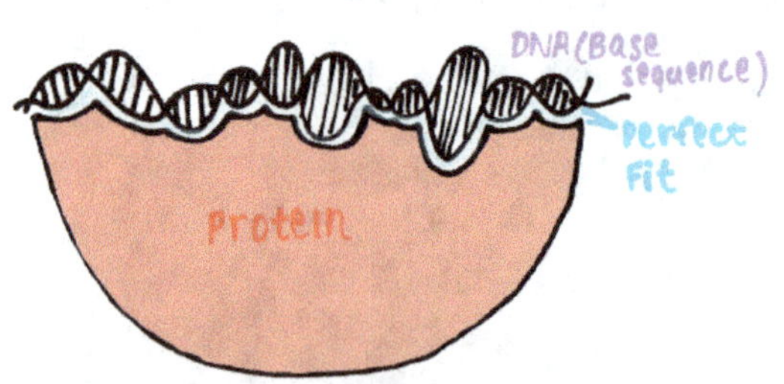

parts of the DNA base sequence of the gene in a perfect fit between the protein and base sequence. And, the activator and repressor proteins determine whether the RNA transcription of the gene will be promoted or prevented. In a unicellular organism, like E.Coli, the changes in the surrounding environment directly affects the control of gene expression.

Differences and Similarities in Unicellular and Multicellular Organisms

In multicellular organisms, gene control does not take place just through the interaction with the environment. The control happens through interaction of the nearby cells. This influences the development of the entire body of the organism.

DNA in a unicellular organism contains only the information required for one cell (itself). The difference between this and multicellular organisms is that multicellulars contain the information required only to define the role of that one particular cell. Every cell has the same DNA, but each cell develops differently because there is only a certain amount of information that is required to be read by that one cell. When it reads its section of the DNA, it is used to synthesize proteins that are needed for that particular cell. In multicellular organisms, gene control is not carried out independently by each cell. Instead, complex processes that involve interactions among cells in the same body occur.

As for the similarities, the ON and OFF switches are what make unicellular and multicellular organisms similar. As we already know, the gene expression is controlled by two types of gene regulatory proteins-- activators (transcription-promoting proteins) and repressors (transcription-preventing proteins). However, in multicellular organisms, the activators and the repressors can be combined to provide a complex type of control (not in unicellular organisms).

Let's look at the differences between unicellular and multicellular organisms' gene controls:

- **Difference 1: The Enhancer**

  Multicellular organisms evolved to form a gene control by allowing multiple gene regulatory proteins to bind to different regulatory sequences along with DNA for a single gene. There is a problem though that occurs-- some proteins might bind to sites, but many base pairs are away from the promoter-- the site where the transcription starts. If it is so far away, how can regulatory proteins turn their genetic switch ON or OFF? This is possible because DNA loops back on itself! That distant regulatory sequence is known as the enhancer. The enhancer is exactly what makes this ability to loop back on itself possible.

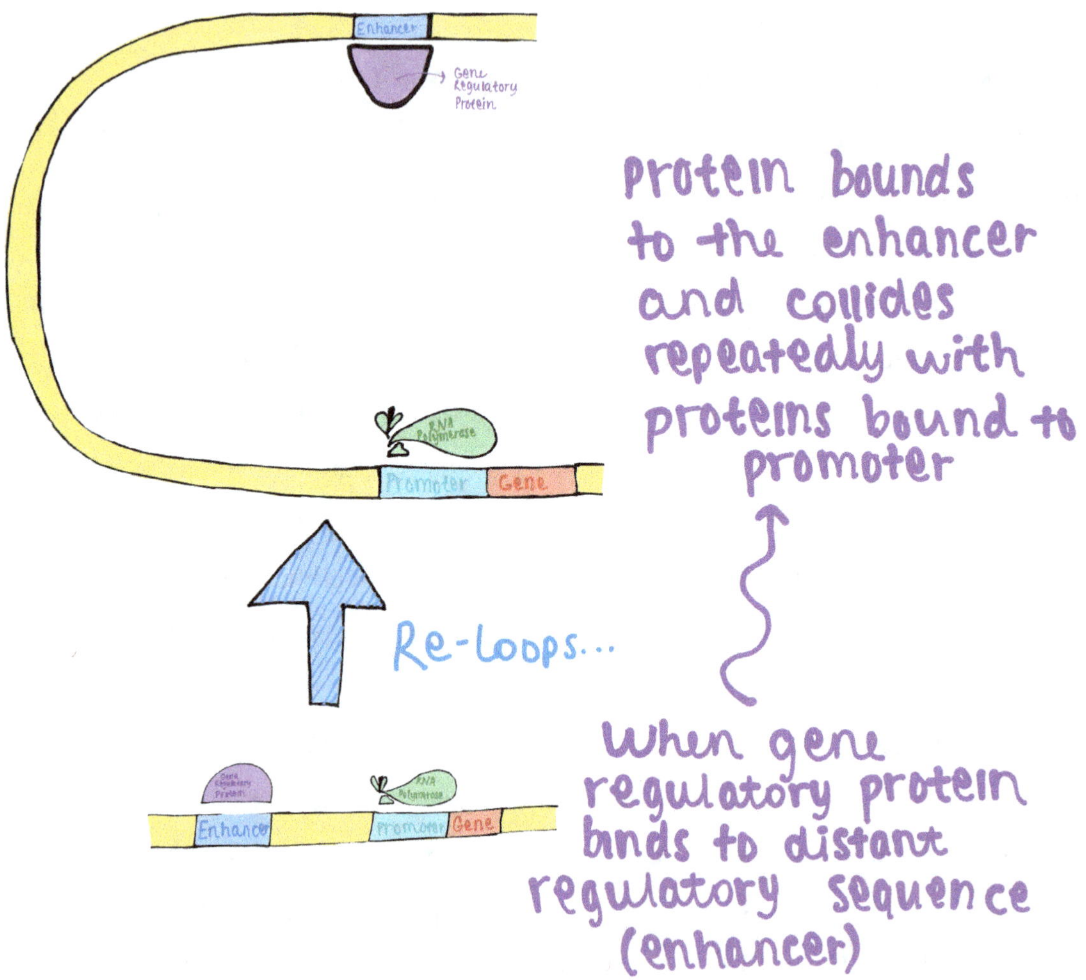

- **Difference 2: The General Transcription Factors**

Gene Regulatory proteins can also speed up and slow down the start of transcription during synthesis. This is due to the fact that the RNA polymerase cannot bind to the promoter by itself. General transcription factors-- a type of protein-- have to bind to the DNA with the RNA polymerase before the transcription may start. The general transcription factors go through several stages in which they bind to DNA. These proteins bind to the T and A bases of the DNA that are found in the promoter. This sequence is known as the TATA sequence. The gene regulatory proteins are involved in that process of binding, getting the control of speeding up and slowing down the rate of the general transcription factors as well as the TATA sequence. Due to this, the gene regulatory proteins may adjust the rate of transcription on their own.

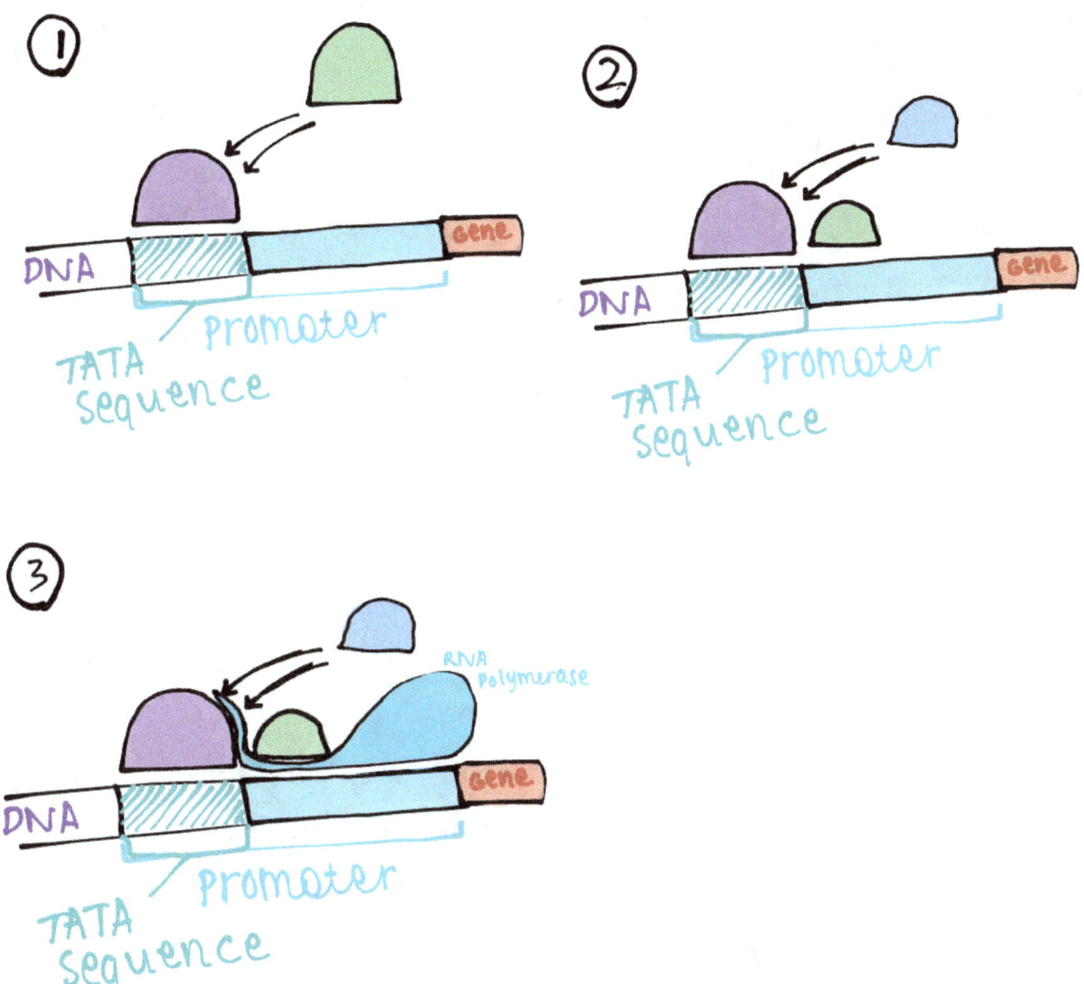

The region where these general transcription factors and TATA sequence happen for gene control takes place in the DNA of multicellular organisms in the gene control region.

The Even-Skipped Gene

Let us use an example of how the enhancers work as well as other different types of proteins work during gene control. As our example, let us use the Drosophila fruit fly again, with its even-skipped gene-- one of the five gene groups (Eve for short). Let's look how the Drosophila's gene in the DNA of one of its cells becomes expressed.

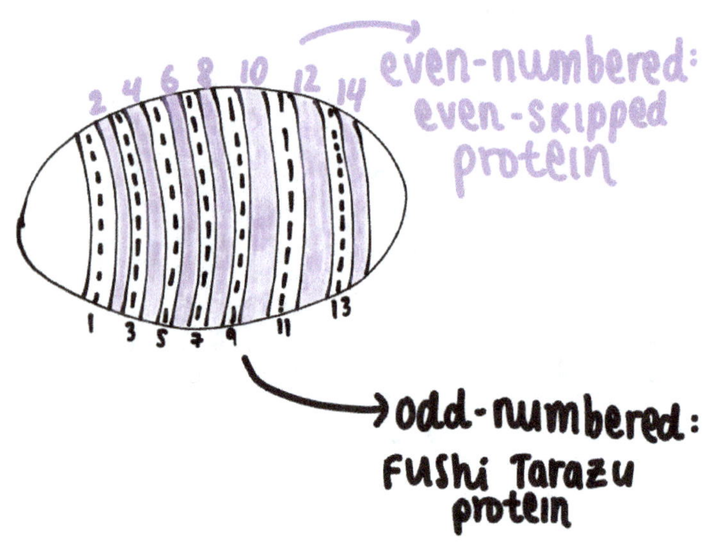

Remember this diagram for the pair rule genes? The pair-rule genes include the Eve gene! *Note: The embryo as shown in the diagram is not actually striped in real life. Actually, the reason for this is that scientists dyed/stained the Eve protein that is produced by the eve gene. "This means that the Eve protein was synthesized only in the parts of the embryo that look like stripes" ("Transitional College of LEX", n.d.). The reason for the Eve gene only being expressed at certain spots is due to the fact that a particular set of gene regulatory proteins appears in those cells, which binds to the DNA there and activates the transcription of the eve gene to synthesize its Eve proteins.*

To see how this works, let's look at only one section-- stripe 2. According to Transitional College of LEX's *What is DNA?*, scientists studied stripe 2 more closely than most other stripes, allowing for researchers and students to look in on that exact stripe:

This is exactly how stripe 2 is expressed-- the combination of presence and absence of the proteins (presence of Bicoid and Hunchback, absence of Kruppel and Giant) causes expression of the Eve gene . Basically, the Hunchback and the Bicoid protein are activators which promote the expression of the Eve gene, whilst Kruppel and Giant are repressors that prevent the eve gene. Gene expression is controlled by a process that is very accurate and hardly ever goes wrong.

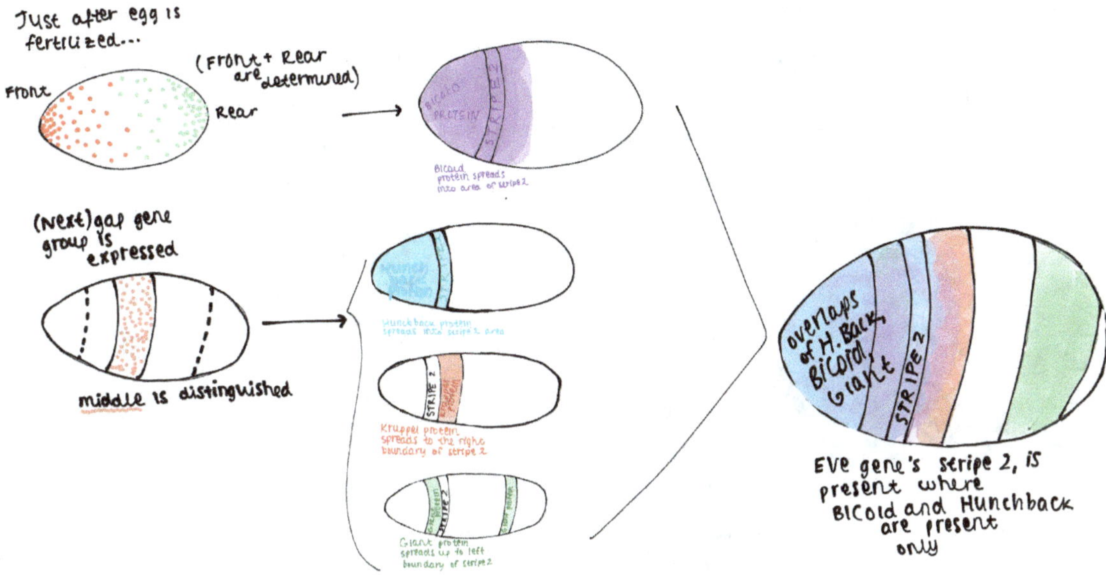

DNA in Stripe 2

Let us look at how the gene regulatory proteins actually turn the switch for the Eve gene ON inside its DNA.

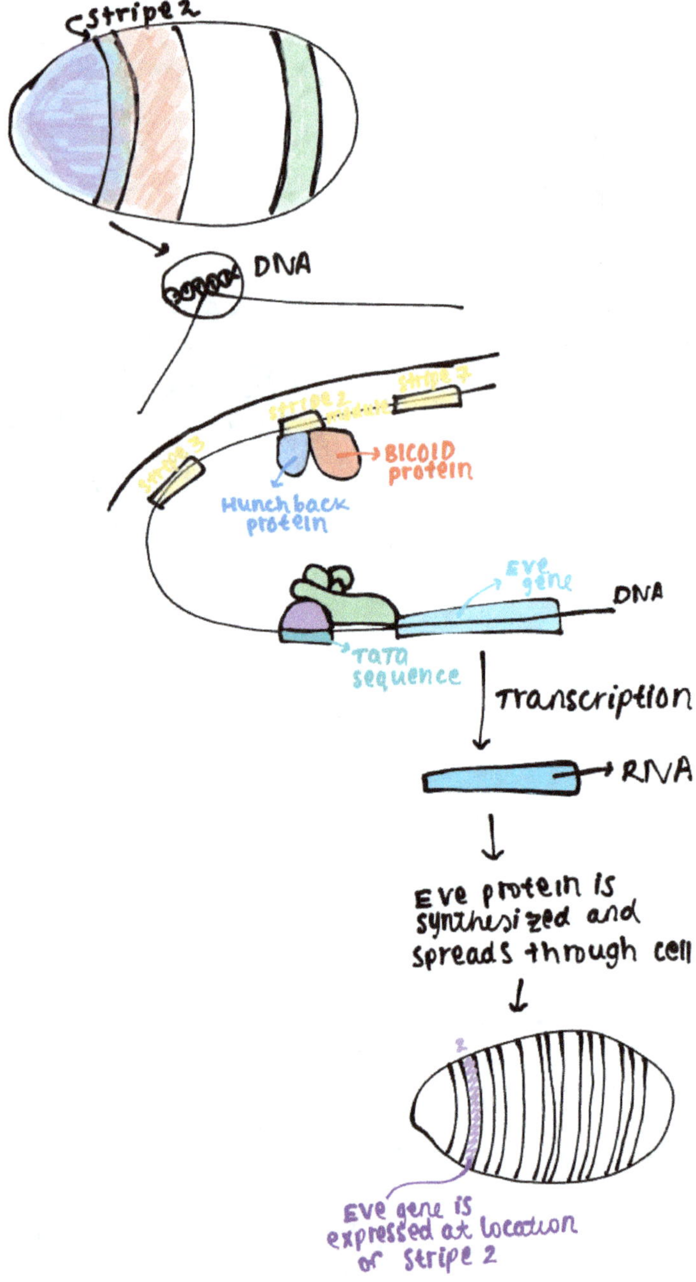

When looking at one cell that is located in Stripe 2, it is seen that the cell contains the Bicoid and Hunchback proteins. Here, the Bicoid and the Hunchback bind to the part of the eve gene control region that controls stripe 2-- this is known as the stripe 2 module. The DNA loops back in on itself to show the start of the transcription of the eve gene. And from there, the eve gene is transcribed, giving the ability to the process of synthesis of the Eve protein.

In the DNA of the Drosophila, the control region for the Eve gene is about 20000 base pairs long, having 20 different types of regulatory proteins bound to it.

Homeotic Selector Genes

The previous genes we discussed are not as complicated as others. Homeotic selector genes are one of the most advanced in that way. Ordinary genes (like the Eve gene) are switched ON and OFF through a certain process, but homeotic selector genes are different. And have a lot of switches. Let's look more into the processes and roles of homeotic selector genes:

- **Pattern of Body Parts:** Up to the expression of the Eve gene, the embryo becomes slowly subdivided into more and more detail-- front and rear, front and back, separate segments, etc. Homeotic selector genes help manage to turn the sections into truly complex body parts like the head, tail, legs, fingers, wings, etc. The expression of these genes produces the specific body parts of multicellular organisms. We previously spoke about what happens when homeotic selector genes become mutated during the development of the Drosophila: if there is a mutation that damages the gene controlling the difference between two different body parts, there might be a problem with them matching up (EX: legs growing out of its head instead of antennas). Mutations like this are extremely important because they are in charge of providing the cells with information on their positions inside the embryo. This is known as the positional information, which creates a structural pattern of the body as it develops. "The most detailed studies that biologists have made of the homeotic selector genes are with the fruit fly. But it turns out that similar genes are found in all animals, including [humans]" (Transitional College of LEX).

- **Gene Sequences that Match the Order of Expression:** Homeotic selector genes is the way that eight of the genes that play a role in forming the pattern of segments in the body of Drosophila, are lined up on DNA's chromosome. They are arranged in almost the same order (nearly) that they are expressed with the body, from front to rear. Other genes, however, do not do this at all. The way genes are arranged on the DNA has nothing to do with the order of expression in the developing organism. However it looks like the homeotic selector genes might be activated, one at a time in a process that advances along the length of the DNA, and the process must advance along the DNA according to some indicator inside each of the cells that show how far it is along the axis of the body of the organism.

- **Big DNA Sequences:** Homeotic selector genes and their control regions take up 650,000 base pairs for their eight genes! That is a lot more compared to the eve gene-- 20,000 base pairs.

- **Storing Positional Information:** The control region of a homeotic selector gene is the entire section of DNA that contains all regulatory sequences that have to do with that gene. That region stores the positional information that comes to be by the proteins that were synthesized in the stages of gene expression.

- **Control Regions:** The control regions of the homeotic selector genes contain binding sites for the proteins that are synthesized by the previously expressed genes like the bicoid gene, hunchback gene, and even-skipped gene. The control regions are able to interpret many parts of information that is provided by the many gene regulatory factors and dictate whether or not to transcribe a certain homeotic selector gene.

Basically, there are five main parts to known about how homeotic selector genes work: Homeotic selector genes are important when forming an organism's patterns that determine the structure of its body parts of the still developing embryo, the sequence in which the gene lie along the chromosome corresponding to the order that the genes are depressed along the body, the coding sections of the homeotic selector genes are everywhere along a sequence of regulatory DNA that is 650,000 base pairs long, the control regions for the homeotic selector genes are like memory chips-- storing information about the position of a cell, and lastly, the control region of a homeotic selector gene is not as simple as going ON and OFF-- instead, it receives many inputs to produce one output (very similar to technology today like microchips).

So, we looked at how DNA works when creating multicellular organisms and unicellular organisms. So, now let us address DNA by using Darwin's theories on: EVOLUTION. Why did animals, including us humans, change? Did DNA affect it? Did we really come from monkeys? Are we all related when going all the way back in time?

Peter stopped. The Lost Boys were searching for him, but he did not dare say a word as they looked around. He knew Tink would find him soon enough, so he continued his reading faster than before. His thoughts were now all over the place. He thought, Does the Tick-Tock Crocodile have this as well if he is technically a beast? Does Hook have this? Does Tinkerbell-- she is a living being, right? Am I like this too? Peter then quickly resumed focus and continued reading, noting the Boys' growing irritation. He had to finish it.

Section 4: Evolution of Plants and Animals

When looking back at how humans supposedly came to how they are now, it is interesting to wonder about how we become so complicated. These days, scientists believe that evolution consists of two parts-- mutation and natural selection. Mutations happen randomly in DNA for many different reasons. Those mutations can be so big that they can change the function of an organism, or its form. Many organisms are born with different mutations, and their mutation ability comes to part when trying to compete against each other for survival. The ones with the 'best' mutation survive. Meaning, the mutations that allowed them to survive the harshest of the environmental conditions. This process is known as natural selection. "The result, say scientists, is the world of creatures we see today, with their incredible variety of forms and sophisticated functions" (College of LEX).

So, if everything about mutation is random, how can they produce changes like eyes, hands, or a brain?

Although scientists have made experiments on the topic of evolution, much is unknown about it, therefore there are many concepts we do not understand yet. The reason for this is that throughout the history of life on Earth, the same exact events would never happen the same again, so it is difficult to do research on. However, to answer this we can look at a popular example of evolution: the giraffe.

The Okapi is theorized to be the main ancestor of the giraffes today. Okapis however had a short neck and a plump body. Although the giraffe clearly has big differences compared to this animal-- especially some parts, like the famous neck of the giraffe. If it is true that the Okapi is the main ancestor that evolutionized into the Giraffe, then there must have been a moment when its neck grew longer. However, if its neck would grow, then it would lose balance and living would be much harder.

Also, since Giraffes have a much longer neck, their heads are much higher from the ground. To pump blood to the head, it would need a stronger heart and its blood pressure would need to be higher. The Giraffe also has valves in the veins of its neck, so the blood would not flood the brain and kill the Giraffe when it is bending down.

So, for the Okapi to grow into a giraffe, it would need to grow a longer neck, longer legs, raise its blood pressure, and get blood flow valves in its neck. Scientists and archeologists have found fossils of Giraffes and fossils of Okapis-- however, no fossil of the animal in its in-between Okapi and Giraffe stage. As a conclusion, scientists decided that: *Evolutionary change does not occur one step at a time, but changes at the same time at once!*

So, how does change happen symmetrically? Meaning, how does one side of an animal not look lop-sided on the other side of the animal?

"When a set of elements is in a symmetrical relationship, a single change-- adding a new axis for example-- can transform the entire set of elements at once. Symmetry is order" (College of LEX). This means that evolution does not actually occur through a random mutation, but with the addition of something different-- everything of an organism changes when its symmetrical order changes.

Now that we know how evolution works, let us look at how and why it happens at the microscopic level.

DNA and Evolution

Evolution really is actually the changes in DNA. Going back to the Okapi-- when the Okapi evolved into the giraffe, the big change in the way it looks is actually caused by the change in its DNA.

Introns and Spacers actually have a lot to do with evolution (see pg 52). Let us look more into the evolution of DNA and what it has to do with spacers and introns.

DNA's Evolution

Early Cells

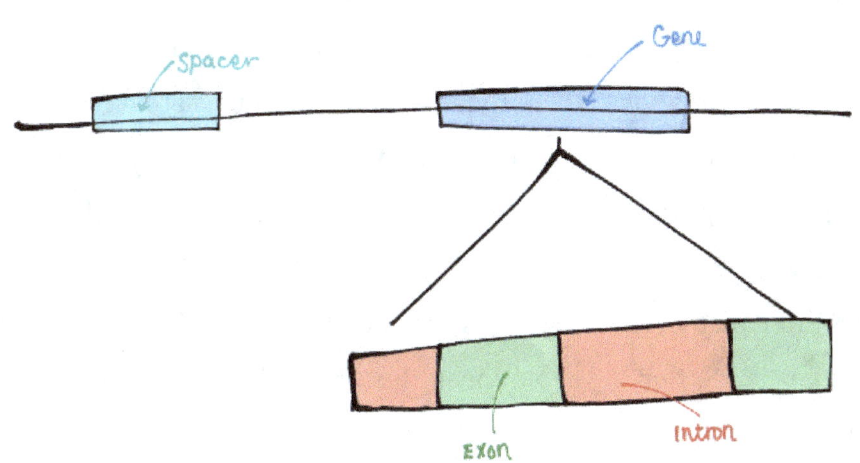

In multicellular organisms, a gene would usually contain exons (used to code proteins) and introns (do not directly code proteins). When protein synthesis happens, the introns are removed from the splicing process that follows then with the RNA transcription. Unfortunately, scientists say that they still do not know what introns are for though. E.Coli, which are prokaryotic and unicellular organisms, do not have introns. They only have exons. Multicellular organisms have tons of introns though.

When looking back at the stages of human evolution, it can be observed that life began with primitive, prokaryotic, unicellular organisms. The next cells that evolved were the eukaryotic cells-- which had a nuclear membrane although they were just single cells. Later, and lastly came the multicellular eukaryotic organisms-- animals and humans!! Due to knowing that prokaryotic unicellular cells like E.Coli had no introns, scientists assume that organisms acquired introns for some apparent reason, while they were evolving. There is a problem to this theory though-- there were cells that were earlier than the prokaryotes… and they had introns.

Now there is a theory that the procaryotes shared a common ancestor with eukaryotic organisms-- the ancestor cell that had introns. Basically, from this ancestor cell, two different types of cells evolved from it. The eucaryotes perpetuated themselves by adding the number of introns in their genes, and the procaryotes just never had the introns in the first place.

DNA repeats itself in many different ways. Its repetition of itself is closely linked to evolution as well!

Globin-Gene Family

There is a certain protein known as the hemoglobin. It carries oxygen through the bloodstream. The hemoglobin is the exact reason why our blood is red. It is composed of two types of globin protein-- the α-globin and the β-globin. Combinations of the α-globin and the β-globin create one globin. The globin proteins that are found in adult mammals are different from the ones that are found in fetuses. In fetuses, globins are known as ε globin, $γ^G$ globin and $γ^A$ globin, which are used to form the β-globin. When those globins become adults, the β-globin and the δ-globin are used.

The genes that code for these many types of globins are known as the globin gene family.

Biologists have made many studies on the evolutionary history of the globin genes. They have found that the amino acid sequences and structure of those genes are homologous-- they are similar in a way that demonstrates a common evolutionary origin. They have evolved from the same ancestral gene, suggesting that the members of this gene family were produced by a process known as the process of gene duplication.

Based on the many differences between the bases in DNA sequences of the genes, scientists are able to estimate when gene duplication produced the branches of the α-globin and the β-globin, which turned out to be about 500 million years ago.

Gene duplication gives organisms the ability to diversify their functions. It is an important part of the idea of evolution, due to having the ability to: preserve the ability of the organism for survival, and yet change it to create changes that make the organism evolve even more. Duplication allows for the genome of the organism to try out different evolutionized forms by maintaining the original function of a gene on one copy while changing it on another one. Basically, the history of evolution is all in the genes. By tracing the growth of the globin gene family, scientists are able to see how evolution works and how other organisms may evolve-- repetition in DNA (a very important part in evolution)!

DNA goes under many types of change besides duplication. One of the most common is known as genetic recombination-- this does not always create evolutionary changes, but it is important to look at due to the fact that it shows how mutations can occur in DNA (thus in our genetic information).

Genetic Recombination

- **1: Recombination during Meiosis**

Meiosis is a special sort of division of cells that takes place when multicellular organisms create their reproductive cells-- egg and sperm. During meiosis, recombination happens only between homologous (corresponding) DNA sequences on two copies of the same chromosome, one of each of the two maternal parents. Basically, genes are exchanged between two homologous chromosomes, but the genes on one chromosome do not change their location.

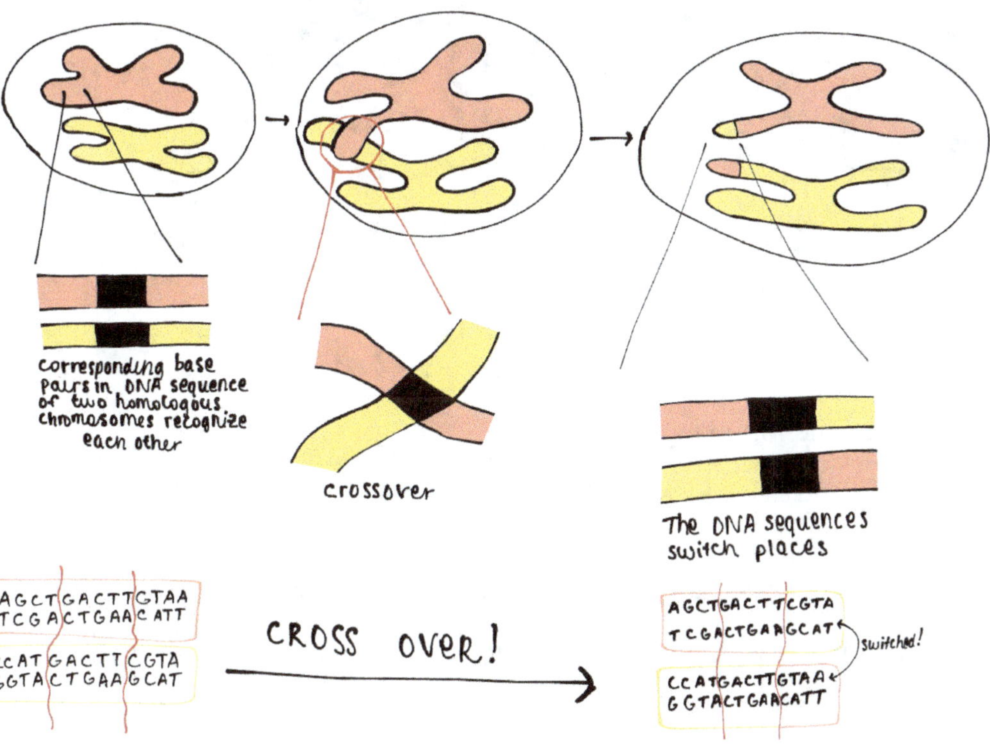

- **2: Recombination through Transposable Elements**

Transposable elements are very short segments-- 100-7000 base pairs of DNA that are sometimes activated to move around to different sites in DNA. This movement of theirs is known as site-specific recombination. Unlike the recombination during meiosis, site-specific recombination does not need homologous DNA. However, it is not known what triggers the movements of transposable elements. Thus, they produce what looks like to be random insertions of case sequences at unpredictable sites in DNA.

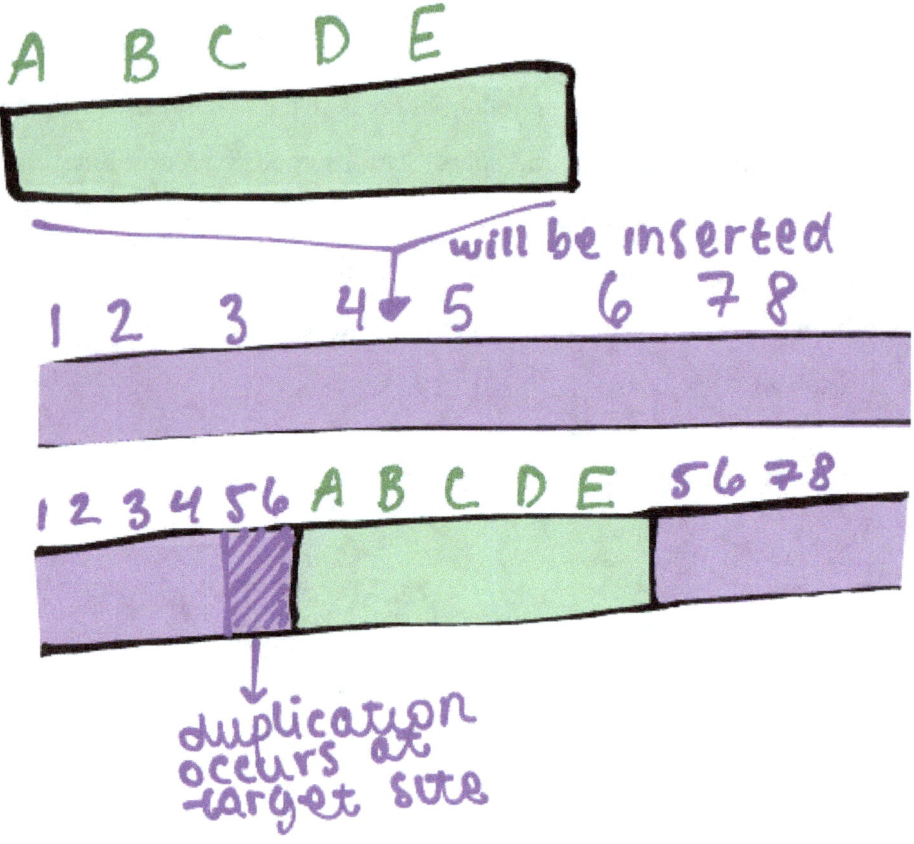

What this diagram is showing are the following two steps:

1. When a transposable element (ABCDE) is put into the target site, it shifts the position of the nearby genes

2. Because of the action of the associated enzymes, the addition of the transposable element always will cause duplication of a short sequence of the DNA at the target site.

Satellite DNA

Chromosomes contain sections where the same short base sequences are repeated over and over again. A series of repetitions like that are known as satellite DNA. Even though scientists performed extensive experiments on the matter, they have not yet figured out what function if there is any that the satellite DNA sequences perform.

According to *The Cell,* "in some mammals a single type of satellite DNA sequence constitutes 10% or more of the DNA and may occupy a whole chromosome arm, so that the cell contains millions of copies of the basic repeated sequence". *The Cell* also speaks about an idea of what satellite DNA base sequences are for:". . . are an extreme form of 'selfish' DA sequences, whose properties ensure their own retention in the genome but which do nothing to help the survival of the cells containing them".

Evolution of Humans

Let us look back at the timeline of how life came on Earth (summary of the introduction and section 3):

The Earth was born 4.6 billion years ago, with the first cells appearing in the primordial soup about 4 billion years ago. The single-cell organism age lasted for quite a long time, with the first cells overcoming environmental obstacles during this time frame. Eventually, through symbiosis, unicellular organisms began to join together to survive together, which led to the birth of the eukaryotic organisms (1.5 billion years ago). About 1 billion years ago, the eukaryotes evolved and became the first multicellular organisms.

Now, let us look at how the stage from the beginning of the multicellular organisms era to the birth of humans.

So, it took 1 billion years from the first multicellular organisms to the first humans. What happened after the first multicellular organisms appeared? They took a quantum leap!!

Once multicellular organisms came into existence, they began diversifying by leaps and bounds. Soon enough, organisms appeared, with many variations in size, shape, form, and functions. But why did this happen? What caused this? Let us analyze the first thing that comes to mind that has always been on Earth-- even today: the environment (specifically, the environment that made this ability of evolution to be)!

An "ever-changing" environment was there during that time. Only organisms that were able to adapt to the environment were the ones that survived, producing offspring and populating their species. This is the process that Charles Darwin called natural selection. All living organisms that are on Earth today, including humans, are what survived during the natural selection process for now. Let us take a look at how organisms survived by adapting to that environment.

Plants and Animals

After multicellular organisms appeared, they split into three main groups: plants (trees, flowers, bushes, etc), animals (dinosaurs, sharks, crocodiles, etc), and fungi (mushrooms, etc). All of the groups were flexible enough to adapt to that environment at the time. However, each of these different types of multicellular organisms adapted and survived in a different way.

Notes: Think about how different plants and animals are-- they actually just used very different ways to adapt to their environment, resulting in the drastic changes between organisms/species. Plants have roots, sprout leaves, have a stem, and produce berries and flowers. And all of them look different-- there are roses, strawberries, pines, etc.

Plants

Plants essentially do not eat food. All they truly need is sunlight, H_2O, CO_2, and through that they perform photosynthesis. Also, plants cannot move around like animals-- they have no arms or legs! So, how are plants similar to animals? They are both multicellular organisms!! Both plants and animals are made up of tons and tons of cells-- animal and plant cells. Here are the differences between plant and animal cells.

Plant cells have a cell wall, and contain chloroplasts and vacuoles. Due to this cell wall, plants are able to grow higher (like trees, vines, etc). Animal cells do not have any of these parts in their

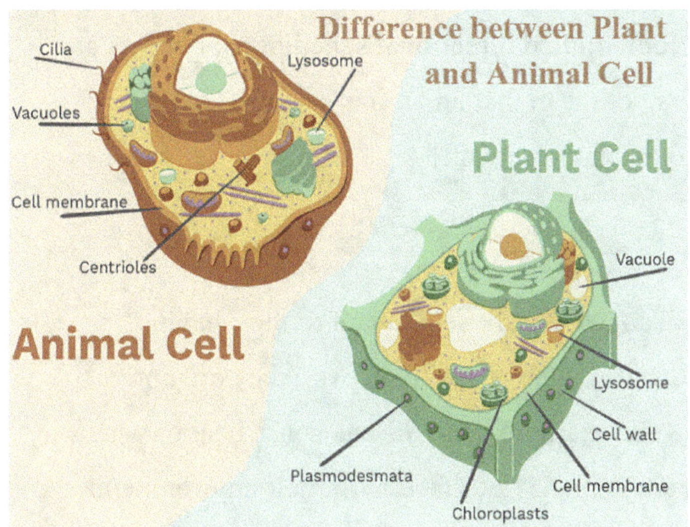

cell. Not only that, but human babies are born nine months after fertilization of their mother's egg, while fruit flies hatch a day after fertilization to be born. Seeds of plants, however, could be hundreds of years old. When a plant is in the seed state, it can remain in that way for a very long time. A century later, they could be planted and could keep on staying that way, putting their development on pause. Animals do not have this seed period, so they cannot stop developing. So, how do you wake up a seed after it has been in 'hibernation' for a hundred years or so? By watering it. When watering a plant, what happens is that the water germinates, giving the embryo the ability to resume development.

At the tip of the root or end of a plant, there are special groups of actively dividing cells that would add more, new tissue to the plant. These cells are known as an apical meristem. The apical meristem produces many cells which later change and become eventually a stem, leaf or root. The process of adding on new plant tissue at the tip of the root is repeated many times, as the plant grows. Basically, the development of a plant may be described as modular (in the type of development that happens). The plant develops sequentially. For the plant to not start growing out of control, the way it wants to, the plant itself has a system in which the different parts of it would stay in constant 'communication' with each other.

Let us use an example from the College of LEX: "if an apical meristem were cut off, but that news was not transmitted to other parts of the plant, the plant would be unable to grow or produce offspring. But the plant's communication system informs the rest of the plant that the meristem has been cut off. What happens next is pretty amazing: another part of the plant, which was not originally an apical meristem, begins functioning as one! As a result, the plant can keep producing new cells and grow" (College of LEX, 270).

Also, a plant's development is very much influenced by the environment. When seeds germinate, it is being influenced by the water that it receives from the environment. When plants grow, they would produce stems and leaves according to the conditions of its habitat. There are many different routes by which the fertilized egg can produce leaves, flowers, roots, etc. When you look at animal development, they follow a predetermined directional schedule that occurs at certain points in time. Although this is so, that does not mean that an animal's adaptation is not affected by the environment. Animals also react to the environment!!

Animals

Animals are not like plants in which they change their shape to adapt to their habitat. Instead, they move or migrate to better environments for themselves. However, they are not absolutely 'immune' to environmental influences-- for example, animals can sense heat or when it's cold through the nervous systems. The nerves are animals' interface with their environment. All animals have a nervous system, not just humans. So do sea creatures, including jellyfish and sea anemones. The nerves are how an animal gets information from the environment it is living in, through the form of light, sound, smell, heat, etc-- for humans, it is the five senses: sight, sound, smell, touch, taste.

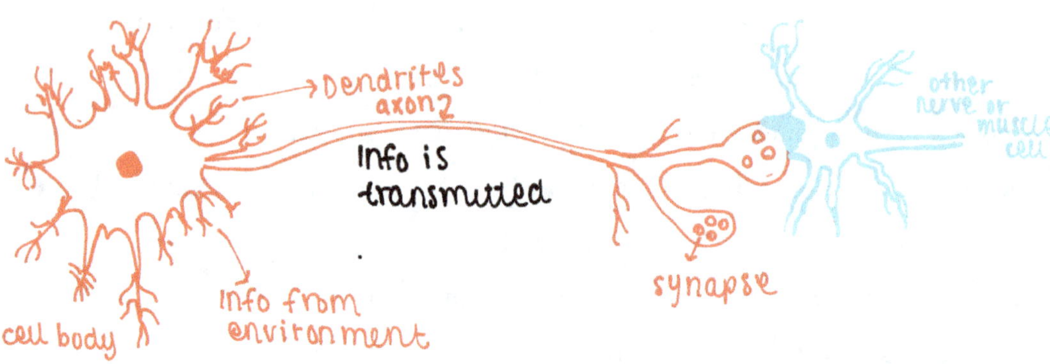

Nerve cells are known as neurons. Neurons receive information from the environment through the dendrites around the cell body-- the part containing the nucleus (the little dot in the middle). Then, the neuron would transmit information coming from the environment as an electric signal over the axon to the branches known as synapses at the opposite end. From the synapses, the signal is transmitted to other neurons or to other cells like the muscle cells. Neurons go under an unusual developmental process. Like all other cells, neurons differentiate and develop from one single fertilized egg, but they do this in a different way from the other cell in the organism's body. The neurons develop in three steps:

- **Step 1:** The nervous system develops from the ectoderm of the embryo. The neurons are created through the cell division. At a certain point, the neurons would though stop dividing. After that, the amount of nerve cells in the body are not able to increase, only decrease.

- **Step 2:** Neurons send out axons and the dendrites, which creates what is known as synaptic connections with other cells. Many of these get sent out. The connections determine how the information flowing through the nervous system will be processed.

- **Step 3:** The system of the synaptic connections is tweaked a bit according to how much electrical signal activity occurs at each synapse. Synapses with many signal activities increase, and less active synapses are taken away or eliminated.

The wiring of the nervous system is determined by information from the environment, even though animals do not alter the shape of their bodies to adapt to the environment like plants. Instead, they compensate by altering the shape of their own nervous system! Once the nerve cells are formed, they stop slicing. Overall, the structure is always being adjusted as it gets new and increasing amounts of information from the environment. The refinement process doesn't stop until the animal will die.

Animals have another system that adapts to the environment -- their immune system. Like the nervous system, immune systems receive information from the environment. In their case, the information does not come through the five senses. Instead, it comes in the form of bacteria and viruses. Through the immune system, our bodies can constantly respond to the information throughout your entire life.

Let us use an example: you get the measles. The immune system goes through the production of antibodies to make sure you will never get the measles ever again. Like the nervous system, the immune system has a memory. However, immunity to measles does not give you immunity to other diseases-- flu, chicken pox, COVID, etc. A certain type of antibody is only effective against a specific disease. This body cannot predict what bacteria or virus will attack it, when it will attack, or where. The immune system can only fight an attacker when it comes into the immune system. Technically, each type of immunity is a specific acquired skill-- fighting a certain virus or bacteria.

There are two different types of immune cells-- B cells and T cells. B cells congregate in the lymph nodes and look for foreign substances that circulate in the body fluids. T cells actively patrol the body by circulating with the body fluids themselves, and attack the foreign substances that they find. Both T cells and B cells have a very VERY short lifespan. A given quantity of B cells is reduced by half within two days of their manufacture. Your body always makes new immune cells to make sure the bacteria and viruses will not make you ill. As we know, there are bad bacteria, but also good. Does the immune system ever attack the wrong substance by accident? The answer is actually very rarely. The task for your immune cells is to distinguish between foreign substances and the ones that belong to its own body. However, although the body cannot afford to attack itself, sometimes it may overreact. Hay fever as well as other allergies occur due to the immune system overreacting to the presence of some foreign substance. That is why our eyes would tear up and our noses would run.

Basically, the nervous and immune systems serve as representatives of the animal body in receiving information from the environment. Animals and plants each have their own way of adapting to their own environment.

Extinction

Fossils have been able to show researchers and scientists of the many different species of multicellular animals and other organisms that have appeared in life on the Earth. In fact, the oldest fossils that have been found, those live organisms, are only about 600 million years old. From then on, there was a "virtual explosion" of life. Most of them became extinct though before the huge changes that transformed the entire ecosystem. In fact, about 99% of all the life-forms that ever appeared on Earth are now extinct!!

There are many theories about what happened to those many life forms. There really is no single cause though. All scientists know for certain is that many species appeared in the past, and all adapted to their environment in their own different ways. Many books are written about the period when many of the extinctions occurred. Geology divides up Earth's history according to fossils found at different points in time. Geologists do not only study rock and soil, but also the study of fossils. On the next page, there is a geologic time chart. It is divided into many ages according to fossils:

ERA	Millions of years ago	GEOLOGIC TIME	
		PERIOD	EPOCH
CENOZOIC		QUATERNARY *Age of Man*	Holocene
			Pleistocene
	2.5	NEOGENE	Pliocene
			Miocene
	23	PALEOGENE	Oligocene
			Eocene
	66		Paleocene
MESOZOIC		CRETACEOUS *Age of Dinosaurs*	Upper
			Lower
	145	JURASSIC *Age of Belemnites*	Upper
			Middle
			Lower
	201	TRIASSIC *Age of Cycads*	Upper
			Middle
	252		Lower
PALEOZOIC		PERMIAN *Age of Amphibians*	Lopingian
			Guadalupian
			Cisuralian
	299	PENNSYLVANIAN *Age of Plants*	Upper
			Middle
			Lower
	320	MISSISSIPPIAN *Age of Crinoids*	Upper
			Middle
	359		Lower
		DEVONIAN *Age of Fishes*	Upper
			Middle
	416		Lower
		SILURIAN *Age of Corals*	Pridoli
			Ludlow
			Wenlock
	443		Llandovery
		ORDOVICIAN *Age of Straight Cephalopods*	Upper
			Middle
	488		Lower
		CAMBRIAN *Age of Trilobites*	Furongian
			Series 3
			Series 2
	542		Terreneuvian

(End of Mesozoic and Beginning of Cenozoic) **Great Extinction #3:** Dinosaurs become extinct

(End of Paleozoic and Beginning of Mesozoic) **Great Extinction #2:**
90-96% of marine life becomes extinct

Multicellular animal fossils appear from this point

The biggest division on the geologic time chart is between the Cambrian Period-- earliest period in which multicellular animal fossils have been found, and the Precambrian (the age before Cambrian). The boundary occurs at the point at about 600 million years ago. The time from that boundary to the present of the 21st century is divided into three eras. Then, each era is divided into periods. Most of the period is named for a region of the world where the fossils from that period have been found (most common place). When many species that had once dominated a given space suddenly died, it made a big, dramatic impact on the ecosystem-- system of relationships among different species of organisms.

Dinosaurs and Mammoths

Whilst on the topic of extinction, let us talk about one of the most famous species of animal that became extinct many centuries ago-- the dinosaur. Their age began 225 million years ago, during the Triassic Period. As we know, about 400 million years ago, the first organisms left the ocean and started to live on land. They were plants. Later, came the first animals-- all of which were reptiles.

The first dinosaurs were very small. There have been discoveries of fossils of the early herbivore dinosaurs, and they were only 80 centimeters long! The Jurassic period was the biggest age for the plants, growing bigger in population and more diverse with new living organisms. There were now cycads, conifers, ferns, and more. All were gymnosperms-- plants that produce seeds in the open. By the middle of the Jurassic period, herbivore dinosaurs reached the pinnacle of their growth and development. During these times, dinosaurs like the long-necked Diplodocus already existed. The biggest dinosaurs like these were 20-30 meters long. To sustain themselves, those dinosaurs would eat all the time. They developed many interesting evolutionary abilities to do that, including teeth that would replace old ones that were worn down, and the presearnes of grinding stones in their gizzards to help with digestion. During this time, a 'revolution' had started in the plant world. By the middle of the Cretaceous Period, gymnosperms were decreasing, and instead, angiosperms-- flowering plants that attracted insects (very colorful as well), began to develop and grow as a population. This was due to the fact that the change of the guard had to do with their efficiency with which the different types of plants could reproduce.

To fertilize each other, gymnosperms had to produce a lot of pollen, then they would wait for the wind to carry it to the pistils of other plants-- remember, insects were not attracted to them, so they had to wait on the environment to adapt-- they could not. Angiosperms were able to produce less pollen, but they had a better efficiency, as the insects that were attracted to them would carry their pollen to the next plant they would travel to. Slowly, gymnosperms disappeared, and were replaced by grasslands and rainforests that were full of angiosperms.

During the Cretaceous, the big lizards would slowly give space to the smaller dinosaurs who were anatomically closer to the ground, so they would eat little plants and bushes that were somewhat too close to the ground. Paleontologist Robert Bakker theorized that the role of the small dinosaurs were to work and change/manipulate the landscape, quickening the disappearance and technical extinction of the gymnosperms from the continents, and replacing them with the fast-growing flowers.

Some scientists thought that the dinosaurs became extinct due to the changes in vegetation-- changing and minimizing what they could eat. By the end of the Mesozoic Era, dinosaurs were completely extinct with the plants that they ate. Small mammals then became the main living beings.

Moving along the years and centuries, mammoths, as we know, are now no longer alive. Mammoths mostly lived in Europe, Asia, and North America during the last 130,000 year of the most recent ice age. 10,000 years ago was when they became extinct. Now, we know that elephants are the relatives of the mammoth.

Mammoths were 3-4 meters high to their shoulder, and had tusks that were about 4.5 meters long, needing to eat 150-200kg of plants every day due to their incapability of efficiency of digesting their eaten food. They were social animals, migrating when needed in groups of 15 or 20.

About 90,000 years ago, a different type of human known as the *Homo sapiens sapiens* came to be. The humans used unpolished stone tools, which as scientists now know were common tools during the Old Stone Age. When surviving in cold climates, they clothed and sheltered themselves with animal hides. For their weapons and other life-styling tools, the *Homo sapiens sapiens* started hunting none other than the wooly mammoths. In Ukraine, scientists found the remains of a 15,000 old dwelling made of mammoth bones-- the builder used the bones of about 95 mammoths!!

11,500 years ago, modern humans came to be, originating in what is known as New Mexico. They made obsidian spearheads known as Clovis points. Although these were implements that belonged to the Old Stone Age, they were very well made, with very sharp tips. Only a thousand years after the invention of the new weapon, mammoths and many other land mammals became extinct from North America.

For mammoths and other animals however, their extinction was not only because of humans. To find the answer, let's look at the environment that the mammoths lived in. 11,000 years ago, the last ice age was still present all over North America. It was not as cold as people might think. Many plants survived very well, and so did mammoths and other large animals. As the ice age ended, the climate went through an abrupt change-- it got much warmer. Summer was very hot and dry. Now plants that mammoths thrived on (that were only able to grow where it was quite cold-- like during the ice age) started growing farther North than where they used to. Mammoths, unable to get an adequate food supply, started to starve.

The Earth's organisms are all connected by eating habits of each animal or plant or fungus-- known as the food chain. Plants depend on the sunlight so they can grow. Then, they are eaten by herbivores, who are then eaten by the carnivores. Not only that, but bacteria in the soil break down then the animal waste and carcasses into more soil, starting the loop again. When thinking of eating as the process of consuming energy, then it can be said that the food chain began with the birth of the first cells. Over the next 4 billion years since their birth, as species became more diverse, only the organisms that found a place in the food chain were able to survive through the years. Curing any part of the food chain creates a lot of damage to the entire chain. When climatic change took away food supply for the mammoths, their food chain began to break and change. This also impacted the appearance of human hunters and the enormous size of the mammoths themselves. It is better to say that, not only did humans cause the mammoths' extinction, but so did the factors that altered the food chain.

Organisms interact with their environment, as we know. Environment selects for the changes in organisms which enhance their ability and probability to survive. This means that life that adapts successfully to its own environment will survive as well as evolve. All in all, this is basically Darwin's concept of natural selection.

From the Apes and Monkey to the Humans

When Darwin published his theory of evolution, many believed it to be a very controversial topic. Especially through Christian theology, which says that all creations were made by G-d, it seemed strange that people had evolved from apes. Scientists have not been able to find any fossils between the two species of man and ape for a long time. So, when did our ancestors known as hominids branch off from the apes through evolution? Or what was an apes' first human-like characteristic?

It has been thought that the first human-like characteristic would be the size of a human's brain, since humans *theoretically* are more intelligent than apes. So, it was assumed by the first researchers that the brain size was the biggest distinction between humans and apes. However, in 1974, there was a famous discovery in Ethiopia, discovering the *Australopithecus afarensis*, the oldest fossil of a hominid as of now, which turned this theory around. The skeleton was named Lucy, and had a brain the size of a chimpanzee's. Her fingers and jaw also were similar to a chimpanzee. The one standing out feature that was different from an ape was the structure of her pelvis-- showing that she walked on two legs. As of now, this is known as the first characteristic that distinguishes hominids from other primates-- the ability to walk upright.

So when did humans start to differentiate from apes? To answer this question, scientists analyzed the entire base sequence of mitochondrial DNA in apes and humans. The conclusion was that Hominids "diverged" from the apes about 5 million years ago. So what happened? According to a theory made by Dr. Yves Coppens, it started with a massive movement of the Earth's crust in Africa. Many earthquakes and volcanic eruptions led to a new environment, with a new range of mountains, extending across Africa from the north to the south. The mountains would block moist winds that blow across Africa from west to east, which cut off rainfall and turned the jungle into a savanna. This was a huge problem for the apes because their food supply was barely sustaining the ability to live in the new environment. Apes were forced to migrate to a different place, trying to find their food in the only forested places left. Apes used all of their four limbs-- arms and legs to move around in the trees, but as there were less trees and more of a dry place with no trees at all, it would be more suitable for standing upright and walking. Scientists believed that apes got into a habit of moving on their hind legs, having no need to climb when crossing the Savanna. Eventually, as their travels became long, the habit of walking stuck to them. Walking upright gave the apes a few advantages, including the ability to carry their food, helping them see farther-- warning of predators, and it would keep them cooler by reducing the area of their body that is exposed to the sun while increasing their exposure to cooling breezes.

Once the apes were able to move across the land on their hind legs, the first hominids underwent a change in how their body works and forms. Now, they developed finger muscles as they were free to move their hands. This helped them develop the ability to use tools. Standing erect also changed and expanded their throat cavity, making it easier for them to produce more complex and vocally louder sounds. Hominids were able to speak, with a language that eventually was able to expand the brain's thinking abilities in different ways.

This ability to speak allowed the hominids to be unique among other animals in their ecosystem. Once the hominids began making tools, their lifestyles changed-- they became very different compared to their near relatives. With tools, humans ". . . being freed themselves from the restriction imposed on the original hominids by the natural ecosystem". For example, during the cold, animals have to adjust by growing fur over themselves. Human beings, however, would make clothes and build heated shelters by making fires. They were able to live in places where their predecessors were not able to survive. Their ranges in the habitat expanded enormously.

Humans could use language as a tool for communicating with others of their same species. They formed groups, communities, and societies. When large groups cooperated, they were able to hunt animals that they were incapable of catching when hunting alone. Later, they were able to create their own vegetation, growing their own food and manipulating the environment for their desires.

Evolutionary Theories

Hugo De Vries

Dutch botanist Hugo DeVries was one of the three scientists who rediscovered Gregor Mendel's laws of inheritance (spoken in Section 1). De Vries performed hybridization experiments with flowers and noticed the appearance of traits that never existed in previous generations. Not only that, but the new traits were passed onto the next generations here onafter. As an example, let us use a red and white flower. According to the laws of heredity, the offspring or a red and a white flower should be either red, white or pink-- a mixture of the two. When De Vries did this experiment, he noticed that one of the offspring was all black-- before a nonexisting trait! De Vries referred to these changes as mutations. This showed that it was possible for new characteristics to appear suddenly, looking as if out of nowhere. This was a huge discovery for the science world. Together-- the concepts of mutation and natural selection-- offered a straightforward explanation on how evolution happens.

Hugo De Vries was born in February 1848 in the Netherlands. He was a botanist and geneticist that introduced the experimental study of evolution. His rediscovery of Gregor Mendel's principles of heredity and the theory of biological mutation is considerably different from today's understanding of the concepts of naturion of variations in species. He was educated at the universities of Leiden, Heidelberg, and Wurzburg. He became a professor at the University of Amsterdam in 1878, and taught there until 1918. In 1886, he performed his red and white primrose experiment. His method of experimentation was different and new, compared to the old method of observation and inference. He was the one that named the idea of change within evolutionary sequences "mutations", showing that they would arise suddenly, as distinct from Darwin's variation of species through natural selection. De Vries believed that varieties are an example of evolution that may be studied experimentally and conceived of evolution as a series of many abrupt changes radical enough to bring new species into existence. De Vries also contributed knowledge and work in plant physiology by osmosis. In 1877, he demonstrated a relation between osmotic pressure and molecular weight of substances in plant cells.

Synthetic Theory of Evolution

Between the 1930s and 1950s, many saw advances in the fields that are associated with evolution-- molecular biology, and in particular, zoology, botany, ecology, population genetics, and paleontology. The research contributed to the development of what is now known as the synthetic theory of evolution. There are two main points in the synthetic theory that shows and explain how evolution works:
- Mutations occurring in the DNA of organisms
- Natural selection of individual organisms born with the mutations

The synthetic theory forms essentially the foundation of evolutionary biology that is known today.

Evolutionary Theory Now

Many people follow and stand in Darwin's footsteps in the way he answered the questions of evolution. However, it is still unknown exactly what evolution is. Recently, evolutionary theorists have taken advantage of the new research done in biology as well as genetics to see a better understanding of the way how evolution really works. Even though nobody claims that evolution is what has been said so far-- meaning everything is in question-- scientists believe that the synthetic theory is the closest idea to the truth yet.

Now, let us look at how today's evolutionary biologists use synthetic theory of evolution to understand how we came to be.

What is Evolution?

The word *evolution* usually is defined as an image of something that is always advancing physically beyond how it was before. However, evolution does not always work like this.

Evolutionary biologists rather define evolution as "any change in the hereditary traits passed from one generation to the next" (College of LEX). Traits that appear in one generation that are not inherited in the next are not evolutionary. Evolution occurs when traits are transmitted to succeeding generations and spreads throughout the population, establishing as a dominant trait. Sometimes, new traits that spread help an organism adapt to the environment. So, organisms may have a trait that would have helped in previous environments as their ancestors lived during a past time, and have new traits that help with today's environment. For example, the human appendix does not really serve any purpose today, yet it continues to be passed on to the next generation, even though it slowly degenerates. Not only that, but not all evolution takes place where we can see them with our eyes. Many evolutionary changes may occur in genes and are transmitted to subsequent generations without any change to the appearance of the eye.

So-- the first part of evolution is mutation-- appearance of organisms with a new trait, one that other organisms in the same population lack. Some examples of these sort mutations could occur when the DNA in a sperm or egg cell is miscopied during replication, or when parts of a chromosome are broken or duplicated. Many mutations produce changes that are not actually good for the organism itself, however, sometimes the resulting wrong DNA sequences would turn out to be an advantage to the organism's survival or its ability to propagate. This trait that was created by the mutation would spread through the population over the course of a few generations and eventually become normal for the entire species, becoming a normal part of the organisms. Once a mutation occurs in an individual organism, how does it exactly spread throughout its species' population then? Scientists are trying to still answer this question. There are a few theories that contribute to scientists' understanding of this, but so far, the most important one is natural selection. Darwin was the first to come up with the concept, even though other theorists refined and repolished Darwin's theory.

Natural Selection

Let us take an example to explain how natural selection works:

- The Galapagos Islands suffered a drought in 1977. From that, much of its plant life was killed, causing a big change in the supply of food, particularly seeds for the *Geospiza fortis,* a native species of the finch bird. Plants that produced small, soft seeds were wiped out from the drought, so the finch was barely able to survive in their changed environment. The only seeds that were left were hard and large seeds. The change in their available food choices changed their diet. By 1978, researchers found a marker increase of the population of finches that have a difference in their appearance-- thicker beaks. This discovery was a perfect example of natural selection, showing that this change was due to this mutation. Here is what was also found from the researchers that discovered this:

- Beak thickness changed from each individual bird. This is known as individual variation

- Individual finches with thicker beaks were able to eat large and hard seeds much more easily, so more of them survived that drought in 1977 and pso produced more offspring with this new installment into the birds' organism. This is known as individual differences in fitness

- Genes would determine the difference in the beak shape, as thick beaked parents tend to produce thick-beaked offspring. This is known as inheritance or heredity.

The natural selection process begins with the presence in this particular population of a genetic mutation which produces a different beak thickness among the individual birds. This type of mutation-based change is known as individual variation. Since the trait is caused by the mutation and it is hereditary, the same trait will be inherited by most of that individual's offspring. If an inherited trait helps an individual survive, the individual is more likely to produce offspring as a result of its survival. In the next generation the offspring with the same trait shall also survive in more numbers than the ones that do not and will produce more offspring of their own. This is now the number of individuals inheriting a trait increases with each generation.

Back to the example of the finches, finches with thinner beaks are less likely to survive, so they produce fewer offspring, who in turn will produce even fewer offspring of their own as time goes on. As the number of finches inheriting this not- so - good trait decreases, fewer and fewer thin-beaked finches will be in the population. As a result, the survival-enhancing trait of the thick beak finches spread through the entire population, as a result of those with thicker beaks being the only ones left. The finches with thicker beaks will have an easier time cracking open their seeds and food, so more of them survived the drought. They produced more offspring who also inherited this trait.

The key to natural selection is, basically, the difference that arises amongst the individual of a species in their ability to produce the surviving offspring. Basically, it all has to do with fitness. Scientists have even created a formula that describes how the fitness of an individual is determined by multiplying two factors together: survival rate (percent of individual of the population who survive at least up to adulthood) and reproductive rate (average number of offspring produced by each individual):

Fitness=Survival Rate x Reproductive Rate

To determine the fitness of individuals in a certain species' population, researchers go out in the field and count every individual of the species. Let us now use rabbits as an example of how fitness works based on one trait:

- Some of the rabbits happen to have one ear shorter than another. Let us look at a rabbit population with the total number of five rabbits that have this trait for their ears, which is genetically determined, meaning that it can be inherited by their rabbit offspring. Of the five rabbits, let us say that three of them are adults-- meaning, old enough to reproduce offspring. That means that the survival rate for the population is $3/5$. Now, suppose the three surviving rabbits produceone, one, and two offspring. Thus, the reproductive rate per individual rabbit is (1+1+2)/3. Multiplying the survival rate by the reproductive rate gives us:

$$[(1+1+2)/3] \text{ x } [3/5]=$$

$$4/3 \text{ x } 3/5= 0.8$$

This means that the fitness of the individual in this rabbit population is 0.8. This means that the rabbits may be expected to produce an average of 0.8 surviving offspring each. So fitness is essentially the measure of how many offspring one generation is capable of reproducing for the next and on generations.

As the formula shows, the higher the survival rate and/or the reproductive rate, the greater its fitness is-- its ability to reproduce offspring. In the real world, having a high fitness rating is really no guarantee that a particular individual will produce lots of offspring. Luck has to do with it, as well as physical health. But what fitness really is is a population of many individuals producing offspring over the next many generations. The larger the population and the more generations scientists look at, the more accurate this measure of fitness shall be. Basically, what is happening is that probability theory is applied to the process of evolution. If you toss a dice only eighteen times, you would not expect the six faces to come up exactly thrice. But when rolling the dice 100,000 times, the number of times each face comes up would turn out pretty even. We can expect our measure of fitness to be quite accurate when looking at a large population over many many generations. Determining the fitness of a species' population turns out is a beneficial way to predict long-term results of evolution for that population of organisms.

Other Theories

So, now we know that evolution arises from differences in the ability of individuals so they produce offspring. The big question that scientists are now asking these days is how did such a simple process produce the amount of variety in the amount of organisms on Earth today, with such interesting and unique/different strategies for survival? Besides the example with the finches, there are more complicated parts of fitness, including when fitness is affected by something else, like interaction of different organisms with each other. Scientists came up with a few theoretical ideas to explain this.

One of the big theoretical ideas is natural selection. Natural selection does not explain everything that has to do with evolution. Random change plays a big role frequently in the process and outcomes of evolution. It is important in the frequency in which a gene appears in a population. This is known as genetic drift. Genetic drift may also be a major factor in evolutionary changes that scientists do not know about yet. There are a few theories across the last century that came up, trying to answer the idea of genetic drift. Here are a few:

- In 1989, American scientist John Cairns performed this experiment:

He studied what happens to E.Coli bacteria placed in a culture with no nutrients-- this is an environment that usually E.Coli cannot survive in. Cairns added lactose, which E.Coli, as we know cannot consume without breakdown. Due to this, the E.Coli are not capable of producing offspring. They would be unable to fo through natural selection. However, among the bacteria, Cairns noticed that the E.Coli had a lactose-breakdown enzyme and began to proliferate. This experiment is probably proving the idea that mutation is directional, meaning it is not random but adaptive and moves in a direction that is for the advantage to the organism's survival.

- Another theory known as the punctuated equilibrium theory was proposed by scientists Stephen Jay Gould and Niel Edlredge.

These two scientists created this theory through investigating fossils. They saw that trilobites-- the animal whose fossil they were looking at-- did not change at all for a long time, until suddenly changing at a certain point in time. It is true that giraffes and bats do this too. There is no gradual or transitional stage between the different versions of the fossils that the two scientists looked at. This was also evidence that mutation is not random. In fact, an organism stays basically the same over a long period of time, and then suddenly ecolces. This phenomenon is known as punctuated equilibrium.

To explain this, let us look at Motoo Kimura's neutral theory of molecular evolution. Kimura investigated the number of mutations in the hemoglobin genes, and found out that the mammals are able to undergo mutations as frequently as about once every two years. This is a more frequent rate of mutation than what was previously thought possible. Based on the data, Kimura developed the idea that mutations occuring in DNA are not selected by natural selection, and instead occur in a neutral manner through the laws of probability. According to Kimura's theory, changes in the DNA that does not affect the individual's survival rate or reproductive rate occur randomly, with an increasing or decreasing frequency without any connection to natural selection. The neutral theory helps explain the idea of DNA mutations, saying that when they appear and are unrelated to natural selection, it highlights the importance of genetics in evolution. Now, the neutral theory is now the main theory of molecular evolution throughout the science of evolution.

We spoke about how mutations work throughout a certain type of species. So, the inherited traits go around at some point to nearly all future generations after many centuries. So, does that mean that all animals of a certain species are related to each other? Technically, yes. If we go far far back, we come from the same path-- humans, other animals, plants, fungi, even unicellular organisms. Together, we evolutionized a new world from the old one. It all started from the beginning of life on Earth, with the environment changing, giving the ability for a different world to exist. A living world. Now, through all information studied and learned, humans were able to complete the biggest task of all:

The Human Genome Project

Peter could not believe what he was reading. Believe it or not, humans are technically related to plants and other animals. Where did he fit in this story? And. . . . where are the fairies in Wendy's story?

Section 5: The Human Genome Project

The Human Genome Project

The Human Genome Project was essentially the pinnacle of Human Biology. It started as an international project in 1990, coordinated by the US Department of Energy and National Institutes of Health. Its goal:

- to identify all genes in the human genome

- to determine the complete sequence of three billion nucleotide base pairs

- Store information in the databases

- Improve the tools for analyzing data and transfer the technologies to the private sector

- Address related ethical, legal, and social issues

The planned timeline was that it would be fifteen years of work, however due to advances in technology, scientists finished the project successfully in 2003.

So, what is a genome?

Genomes consist of all DNA in one organism. So, DNA is the functional units of heredity, determining the characteristic of an organism, with Base pairs that are made of nucleotides. Genes contain the information necessary for an organism to grow, metabolize, reproduce, and carry out other necessary functions of life for the organisms. Individuals may have all the same genes, but may have different forms of genes. The sequence of nucleotides within a gene determines the composition and order of the amino acids that will be produced. Besides coding genes, genomes contain noncoding sequences-- some with functions and some with no function at all-- as we know. The number of chromosomes, the number of genes, and the total number of nucleotides and sequence of nucleotides are characteristics used to describe the genome in an organism. The human genome contains twenty-four chromosomes that are Sex chromosomes-- meaning they are X or Y chromosomes, twenty-two other chromosomes known as autosomes, in which one side fo the chromosome is from the biological father, and the other from the biological mother, and two sex hromosomes that determine the sex and germ cell development (females have two X chromosomes and males have one X and one Y). The amount of genes changes based on the species, and there are rare cases of having an extra sex cell.

The methods that were performed to complete the Human Genome Project were as follows:

- **Sequence the human genome:** There were many complications and technological challenges. However, there was a sudden spike of advancement of technology in bioinformatics, which is through the use of computers that scientists are able to acquire, organize, and analyze biological information.
- The main method that was used is known as whole-genome shotgun sequencing, which was developed by J. Craig Venter from the Institute for Genome Research. This institute was utilized in 1994 by Venter with microbiologist Hamilton O. Smith to sequence a bacterium that causes ear and respiratory diseases and infections.

The twenty-four human chromosomes first had to be broken into smaller pieces, ranging in size. The fragments would then be subcloned into plasmids maintained in bacteria. The sequencing of them works by using the cloned fragments as substances in biochemica. Reactions that generate sets of fragments that differ in length by a single nucleotide. Then, gel electrophoresis would separate the fragments based on their difference in sizes and dyes (the fluorescent dyes would indicate the nature of the last nucleotide base of each fragment that is synthesized during the sequencing reactions. The order of colored bonds (yellow, blue, green, and red) on finished gel represents the sequence of nucleotides in chromosomal fragments. A chromosomal fragment is the size of electrophoretic separation-- about 500-700 bases. Computers then would compile sequenced fragments by finding the overlapping regions and assemble them to form a linear sequence. Twenty four chromosomes ranged in length from 50-25 million nucleotides each. The first draft that was compiled and completed was in June 2000. But the final draft with good quality was finished in 2003.

In May 2006, the last of even more detailed sequences were completed. In May 2021, the next step after the Human Genome Project was completed. The goal was to examine the three billion nucleotides to identify all genes that encode proteins. Computers would search the genome for open reading frames-- the ones that could potentially encode proteins based on the presence of specific sequences known to be necessary for initiation and termination of translation. In the end, though complications with eukaryotic genomes, newer software programs were able to utilize additional strategies to increase effectiveness of the process.

Conclusion

Peter finished the book. As he closed the pages, he eased his tense back and relaxed on the branch he was sitting on. He felt strange. He was not sad, yet not happy. He was surprised that he could understand what the book was about.

Tinkerbell finally found him. She crossly stared at him, angry that he was ignoring her.

Peter: Tink, have I grown up? Why. . . How can I understand what the grownups say?

Tinkerbell: I do not know, Peter. But you do not have to grow up to understand something that is about yourself.

Peter, pointing out to the horizon of Neverland: What do you mean? This book is about the people and animals and plants that live down there.

Tinkerbell: No. None of us here-- the Lost Boys, Hook and his crew– you guys once did come from that world. This does work for you, and there is no shame in understanding it. You may deny it, but this "grown-up" scientific reasoning is a part of you.

Peter: That is life, isn't it? Just, science?

Tinkerbell: You may think of it that way. But think about it, everything that this biological science has created-- plants, animals, fungi-- is beautiful. The trees were created by these processes, so are cats and dogs, even crocodiles. Trees help us breathe, flowers that smell nice and make us feel happier about the world, animals that we love to cuddle with and pets, all of it is a creation through this biology. You were created by this as well. There is no shame in it. You are all a part of this world, are you not? Life that has been created will serve some purpose and that is true. Why else would these trees be created? This science is just an explanation of how the world works in nature.

Peter: I see. I see. Well, I am still Peter Pan and I am still in Neverland. I will never grow up and I will always enjoy life. And now, all of the life that is around me.

A Last Word from the Author

So, what is life? Now we know: it is the breathing organisms that have lived on this Earth throughout Earth's history. Life disappears and new life appears, trying to fight for its survival. What is left is who survived. The ones who made it to the top. The ones who perhaps destroyed others lives for their own were the ones that survived.

We read about Peter Pan and we read about Dr. Frankenstein. Although we do not know yet how it is possible to never grow up (because, doesn't that mean then that DNA stops developing?) and we do not know how to resurrect the dead, we have a foundation of information of exactly how life works, and how it might have come to be.

There is still much to learn in the fields of biology. Perhaps one day we will truly find the way to stop death or find a cure to cancer as well as other diseases.

All living things, whether it is a human or not, are all unique. Their lives depend on others that have survived with them throughout their history. Life has been dependent on other life since the beginning of time. Together, we are able to understand the fundamentals of survival as well as what we are made of.

Biology can help understand many things but there are still theories that make you wonder, *How? How do the cells know what to do? How do humans know what to do?* Basically, what do we think? How do our cells think? What are our instincts? How does our brain work? What exactly is our conscience? Ask yourself that. What are we without having the ability to speak for ourselves and be able to think? How are we able to do this?

Glossary

Adenine: one of the nucleobases of DNA (corresponds with Thymine)

Adenosine Triphosphate (ATP): A source of energy on the cellular level

Alkali: a chemical compound that dissolves in water

Amino acid: molecules that combine together to create proteins

Antiparallel: Parallel, but moving in opposite directions

Atomic spectra: frequencies of radiation of an atom

Autocatalytic systems: certain types of molecules and the chemical reactions between them

Blastula: produced during development of the embryo by repeated cleavage of a fertilized egg

Capsules: utmost structures of bacterial and fungal cells

Carbon: a very common chemical element

Carbon Dioxide: Acidic colorless gas

Cell: smallest unit factor that can live on its own

Cell division: the process of the cell duplicating its contents inside, and then dividing itself to create two cells

Codons: a trinucleotide sequence of DNA or RNA which corresponds to a certain amino acid

Cyanobacteria: a phylum of bacteria that create energy from photosynthesis

Cytosine: one of the nucleobases of DNA (corresponds with Guanine)

DNA: Deoxyribonucleic Acid, holds our genetic information and is made of the double-helix, and four nucleotide bases

Egg-polarity genes: one of the five gene groups spoken in Section 3

Energy flow: Flow of energy through living things

Enzyme: almost always a protein, speeds up rate of chemical reactions in a cell

Eukaryotes: an organism that contains a cell or cells with a distinct nucleus

Exons: a sequence of DNA or RNA molecule containing information that codes for a protein or peptide sequence

Extinction: the dying out, examination of a species

Fushi-Tarazu: a type of pair-rule gene

Gap genes: one of the five gene groups spoken in Section 3

Gastrulation: early developmental process, in which an embryo reorganizes into multilayered structure

Gemmules: a cluster of embryonic cells

Genes: functional unit of heredity

Genetic Recombination: rearrangement of DNA sequences

Glycolysis: a pathway that breaks glucose into two three-carbon compounds

Guanine: one of the nucleobases of DNA (corresponds with Cytosine)

Histochemical: distribution of chemical contents of tissues through the means of stains, indicators, etc

Homeotic Selector genes: one of the five gene groups spoken in Section 3

Homunculus: a representation of a small human living inside the sperm cell that determines children's appearance

Hydrogen: a very common chemical element

Introns: a segment of a DNA or RNA molecule that does not code proteins

Karyomitosis: mitosis

Lipids: a type of macro biomolecule

Lymphocytes: a type of white blood cell

Mammary gland cell: the mammary gland is an organ. The mammary gland cell is its cell

Methane: simplest hydrocarbon, gas that is found in the atmosphere

Mitosen: part of the cell cycle

Molecules: atoms bonded together

Monographs: specialist work of writing on one subject

Morphogenetic: biological process that causes cells/tissues to develop shape/form

NADH: Part of the citric acid cycle (See section 2-3)

NADPH: Part of the citric acid cycle (See section 2-3)

Negative entropy: something becoming less disordered

Nucleic acids: a class of macromolecules found in cells and viruses

Nuclein: a group of proteins containing phosphorus

Oligonucleotides: a polynucleotide that contains small snippets of DNA and RNA

Organs: In bodies, a collection of tissues that form functional unit

Pair-rule genes: one of the five gene groups spoken in Section 3

Pepsin: a stomach enzyme that serves to digest proteins found in ingested food

Phosphate: a form of acid that contains phosphorus, found in bones and teeth

Phospholipids: a group of polar lipids

Polymerase: an enzyme that synthesizes long chains of polymers or nucleic acids

Polymers: a structure that contains large molecules/macromolecules

Polynucleotides: combination of nucleotide monomers connected to each other through covalent bonds

Polypeptides: organic polymer with a large number of amino acids

Preformationism: theory of inheritance (see section 1)

Prokaryotes: living organisms with no distinct nucleus

Prokaryotic cells: a cell with no distinct nucleus

Protein: made of amino acids, contain

Pyrimidines: EX: cytosine, Thymine, Uracil

Pyruvate: output of the metabolism of glucose

RNA: ribonucleic acid, has one helix

RNA polymerase: enzyme that copies DNA sequence into RNA sequence

Satellite DNA: highly repetitive DNA

Segment-polarity genes: one of the five gene groups spoken in Section 3

Splicing: join or insert

T4 phage: a species of bacteria

Thymine: one of the nucleobases of DNA (corresponds with Adenine)

Unit factors: numerical quantity factor

Acetic acid: a type of synthetic acid

Adipose tissue: body fat

Aminoacyl-tRNA synthetase: responsible for performing first steps of protein synthesis

Anteroposterior axis: one of the two sides in an embryo, determines the head of the organism

Anticodons: a trinucleotide sequence

Bilayer membrane: a double cell membrane

Carboxylic acid: class of organic compounds

Catalysis: responsponsible of modifying time of chemical reactions

Catalysts: the ones that modify the time of chemical reactions

Chromic acid: a type of mixture

Citric Acid Cycle: see section 3

Cortex: outer part of an organ

Deoxyribose sugar: pentose sugar

Dorsoventral axis: determines belly or back of the embryo

double helix structure: structure of DNA that holds genetic information

Ectoderm: outermost layer of cells

Flagella: appendage that produces many microorganisms

Genetics: studying genes and genetic information

Glacial acetic acids: water-free acetic acid

Heredity: passing down characteristics/traits to next generation

Leukocytes: a colorless cell that circulates the body

Lipid Droplets: storage organelles

major grooves: occurs in double helix structure when backbone falls apart

Meiosis: formation of egg and sperm cells

Mesoderm: germ layer that rises during gastrulation

metabolic pathways: a series of chemical reactions in a cell

Minor grooves: when the backbones of DNA are close together

Mutation: abnormal change in heredity traits

Nucleotide: building base of nucleic acids

Nucleotide bases: Bases in DNA that store genetic information, the four that there are in DNA are Adenine, Thymine, Cytosine, and Guanine

Peptide bonds: linking of two amino acids

Polypeptide chain: a long chain of amino acids

Polysaccharides: long chains of sugar

protein synthesis: Process of synthesizing protein

Proton-motive force: natural substance found in some foods

Purines: natural substance found in foods

Ribosomes: a minute particle that consists of RNA

Sequential induction: special process for developing cells

Side chains: a chemical group that is attached to a core part of molecule

Spindle fiber: a protein structure that divides genetic material in a cell

Symbiosis: interaction between two different organisms living in close physical association

Synthesis: production of chemical compounds

Tetrahymena: a unicellular eukaryote

Thromboplastic proteins: mixture of phospholipids and tissue factors

Uracil bases: one of the base pairs of RNA that is not present in DNA, corresponds to Adenine

Vegetal pole: one of the two sides in an embryo, determines the tail of the organism

Vesicles: a certain structure inside of the cell

References

"1879: Mitosis observed." National Human Genome Research Institute, www.genome.gov/25520234/online-education-kit-1879-mitosis-observed.

"Ancient DNA helps scientists study human evolution: 'It's like a time capsule'." uchicagonews, University of Chicago, news.uchicago.edu/story/ancient-dna-helps-scientists-study-human-evolution-its-time-capsule.

"Cell Theory." Society, National Geographic, www.nationalgeographic.org/encyclopedia/cell-theory/.

"Deoxyribonucleic Acid (DNA) Fact Sheet." National Human Genome Research Institute, National Human Genome Research Institute, www.genome.gov/about-genomics/fact-sheets/Deoxyribonucleic-Acid-Fact-Sheet.

"Erwin Chargaff." Historica Wiki, Historica, historica.fandom.com/wiki/Erwin_Chargaff.

"Erwin Chargaff." Yourdictionary, Yourdictionary, biography.yourdictionary.com/erwin-chargaff.

"Erwin Schrödinger." The Nobel Prize, The Nobel Prize, www.nobelprize.org/prizes/physics/1933/schrodinger/biographical/.

"Francois Jacob." Biography y Vidas, www.biografiasyvidas.com/biografia/j/jacob_francois.htm.

"Jacques Monod." The Nobel Prize, The Nobel Prize, www.nobelprize.org/prizes/medicine/1965/monod/biographical/.

"Jacques Monod." Wikipedia, Wikipedia, en.wikipedia.org/wiki/Jacques_Monod.

"Mulder, Gerardus Johannes." Cengage, encyclopedia.com, www.encyclopedia.com/people/history/historians-miscellaneous-biographies/gerardus-johannes-mulder.

"Nucleotides." National Cancer Institute, National Cancer Institute, www.cancer.gov/publications/dictionaries/genetics-dictionary/def/nucleotide. Accessed 30 Oct. 2021.

"Robert Hooke." Biography: History&Culture, Biography, www.biography.com/scholar/robert-hooke.

"Searching for protein composition and function." Dr. Marti, bioinformaticshome.com/bioinformatics_tutorials/history/history_Page_5.html.

"Stanley Miller's Experiment: Sparking the Building Blocks of Life." Meteorites &Life, PBS, www.pbs.org/exploringspace/meteorites/murchison/page5.html.

"Stanley Miller, 77 chemist was a pioneer in studying the origins of life." Los Angeles TImes, www.ncbi.nlm.nih.gov/books/NBK21514/.

"Walther Flemming." Wikipedia, Wikipedia, en.wikipedia.org/wiki/Walther_Flemming.

"Watson and Crick." History, BBC, www.bbc.co.uk/history/historic_figures/watson_and_crick.shtml. Accessed 5 Jan. 2022.

, et al Molecular Cell Biology. Vol. 4, W.H. Freeman and Company, 2000, www.ncbi.nlm.nih.gov/books/NBK21514/.

August Weismann picture. "August Weismann." Wikipedia, Wikipedia, en.wikipedia.org/wiki/August_Weismann. Accessed 7 Oct. 2021.

Barras, Colin. "The father of all men is 340,000 years old." NewScientist, Life,
www.newscientist.com/article/dn23240-the-father-of-all-men-is-340000-years-old/.

Barrie, J. M. Peter Pan.

Cullen, Katherine. Encyclopedia of Life Science. Vol. 1, Facts on File, 2009.

Cullen, Katherine. Encyclopedia of Life Science. Vol. 2, Facts on File, 2009.

Dam, Ralf . "Friedrich Miescher and the discovery of DNA." Science Direct, Science Direct,
Friedrich Miescher and the discovery of DNA.

Danny. Robert Hooke: The Genius Newton Tried To Erase From History. Exploring History,
medium.com/exploring-history/robert-hooke-the-genius-newton-tried-to-erase-from-history-794
2427752b0.

Forterre, Patrick, et al. "Origin and Evolution of DNA and DNA Replication Machineries."
NCBI, Madame Curie Bioscience Database, www.ncbi.nlm.nih.gov/books/NBK6360/.

Geological time source. "Rocks and Minerals." Rocks and Minerals,
www.rocksandminerals.com/geotime/geotime.htm.

Goyal, Shikha. What is the difference between plant and animal cell?. "What is the difference
between Animal and Plant Cell?" by Shikha Goyal. Josh,
www.jagranjosh.com/general-knowledge/difference-between-animal-and-plant-cell-1618491661
-1.

Gregor Mendel Picture. "Gregor Mendel." Wikipedia, Wikipedia,
simple.wikipedia.org/wiki/Gregor_Mendel. Accessed 7 Oct. 2021.

Hershberger, Scott. "Humans Are All More Closely Related Than We Commonly Think."
Scientific American, Scientific American,
www.scientificamerican.com/article/humans-are-all-more-closely-related-than-we-commonly-thi
nk/.

Hugo De Vries. "Wikipedia." Wikipedia, en.wikipedia.org/wiki/Hugo_de_Vries.

Learn, Joshua. "No, a Mitochondrial Eve" Is Not the First Female in a Species." Science,
Science, No, a Mitochondrial "Eve" Is Not the First Female in a Species.

Miescher picture. Google,
www.google.com/search?q=miescher+young&tbm=isch&ved=2ahUKEwjtxaejteTzAhUByDgK
HW-3A4AQ2-cCegQIABAA&oq=miescher+young&gs_lcp=CgNpbWcQAzoFCAAQgAQ6BA
gAEEM6BAgAEB5Qx-4BWKv7AWDs-wFoAHAAeACAAW-IAYkEkgEDNi4xmAEAoAEBq
gELZ3dzLXdpei1pbWfAAQE&sclient=img&ei=Ggl2Ye3QI4GQ4wHv7o6ACA&bih=862&biw
=1517&rlz=1CATRYQ_enUS884&safe=active&ssui=on&surl=1#imgrc=xk-62-jnlbgAyM.
Accessed 24 Oct. 2021.

Okapi. Wikipedia,
commons.wikimedia.org/wiki/File:Okapi_(Okapia_johnstoni)_2009-04-04_02.jpg.

Paweletz, Neidhard. "Walther Flemming: pioneer of mitosis research." Scitable,
www.nature.com/scitable/content/Walther-Flemming-pioneer-of-mitosis-research-12650/.

Peter Pan standing outside of the window. *Pinterest*,
www.google.com/search?q=peter+pan+sitting+painting&rlz=1CATRYQ_enUS884&source=lnm
s&tbm=isch&sa=X&ved=2ahUKEwi6gOK4h6H1AhX6JTQIHccsCGwQ_AUoAXoECAEQAw
&cshid=1641607255254544&biw=1517&bih=862&dpr=0.9&safe=active&ssui=on&surl=1#img
rc=TDkMSQs9mlPQHM&imgdii=wQTRG1bVeSo-4M. Accessed 7 Jan. 2021.

Planck, Mark. "Friedrich Miescher and the discovery of DNA." ScienceDirect, 24 Dec. 2004,

Pray, Leslie A.
"https://www.nature.com/scitable/topicpage/discovery-of-dna-structure-and-function-watson-397
/." Nature Education, Scitable, 2008,
www.nature.com/scitable/topicpage/discovery-of-dna-structure-and-function-watson-397/.

Slezak, Micheal . "Found: closest link to Eve, our universal ancestor." New Scientist, New
Scientist,
www.newscientist.com/article/mg22429904-500-found-closest-link-to-eve-our-universal-ancesto
r/.

The Editors of Encyclopaedia Britannica. "Hugo de Vries." Britannica, Britannica,
www.britannica.com/biography/Hugo-de-Vries.

Transitional College of LEX. What is DNA? LRF.

Watson, James. DNA: The Secret of Life. Edited by Andrew Berry, Alfred A. Kopf.

Wong, Sam. "Gregor Mendel." NewScientist, New Scientist,
www.newscientist.com/people/gregor-mendel/.

www.sciencedirect.com/science/article/pii/S0012160604008231?via%3Dihub.

Acknowledgments

I would like to thank Ms. Ogden for making me write this book. I would also like to thank everyone who edited this book, including: Nicholas Shatalov, Sidharth Mathilakath, Audrey Liang, Madeline Ko, and Lily Zheng-- author of *Knowing and Understanding MBTI, Enneagram, and Socionics.* I would like to thank my family for supporting me and all my friends and classmates.

About the Author

Michelle Milman has enjoyed writing for most of her life. She was born in Portland, Oregon, and still lives in Oregon to this day. Michelle is a student at the International School of Beaverton in Aloha, OR. She lives with her parents, her brother Asher, and her stuffed animals. In her spare time, she can be found playing the piano, practicing tennis, drawing blimps and elephants with extremely long legs, and solving math equations. *If Peter Pan Knew DNA* is Michelle's first ever published book- but definitely not her last :)